Reflections through the Mist

Pat Lyne

ISBN 0 9509674-3-2

Published and distributed from
Combe Cottage, Presteigne, Powys. LD8 2LH

A CIP record for this book is available from the British Library.

Limited Edition

Printed by Orphans Press Ltd., Leominster, Herefordshire HR6 8JT.

PREFACE

I have been encouraged to write Reflections through the Mist by readers and by their generous appreciation of Shrouded in Mist and Out of the Mist. The celebration of the seventy year life of the Connemara Pony Breeders Society, 1923-1993, provides me with the incentive to pay tribute to many people and ponies who have been part of the remarkable progress of the Connemara Breed.

It is twenty years since I set out on my pilgrimage, following the Connemara Pony through the Mists of time. It has provided an inexhaustable subject and I am happy to share further discoveries, thoughts and pleasures, which I have savoured on my journey.

My Reflections are written as a sequel to the earlier volumes and for this reason reference numbers are included in order that readers may easily locate a pony or an occasion from previous chapters. I have indulged myself by omitting pedigrees and stud book numbers from the text. To compensate, a comprehensive Appendix records the sire and dam of every pony mentioned.

My thanks are due to many who have supported my work; to Jo Donnelly who has been a faithful copy editor and endless source of encouragement; to Sister Clare for assistance with all the art work and Sister Patricia for the title script; to the Connacht Tribune for permission to reproduce photographs taken at Clifden show from 1954 through to 1993; to many who lent me their personal photographs, with special thanks to Anne Harries and Beatrice Milleder; to Tim Robinson's "Introduction to and Gazeteer of Connemara" which has been my constant source of reference. His knowledge of the region has filled many personal gaps, especially with regard to Irish place names and their interpretation. Finally, my grateful thanks to Peter Patrick Hemphill for agreeing to write the Foreword.

Pat Lyne,

Combe Cottage, 1993.

ACKNOWLEDGMENTS

My grateful thanks are due to the following who have kindly allowed me to use their photographs; the Connacht Tribune, the Connemara Pony Breeders stud book, the German and Swedish archives; the Professional photographers, Carol Gilson, Event Print, Paul Saville-Pleasure Prints, N. B. Jones and Tracy Reep Elliot from the U.K.; Mary Tarry and Joseph Brady from Ireland; Christine Tchir and Ed Moore from the U.S.A., Henning Vibak (Finland), Kees Vis (Holland), Helen Sloane (Aust), Barbara Thompson (N.Z.) and from their private collections; Lord Revelstoke, Mrs. Lynch, Brigadier Berridge, Lesley Llanigan O'Keefe, Marie Luise Schwartz, Nina Gustafson, Frank Glenister, Nicholas Palmer and many others from all corners of the earth. And also to Tim Robinson for permission to reproduce his Bilingual Map of Connemara.

FOREWORD

By

THE LORD HEMPHILL

The author, having been involved with her local Hunt "The Enfield Chase", the Pony Club, and in the production of Hunters and Show hacks and, later, having been invited to join The British Show Hack and Cob Association panel of judges, came in 1961 to own a three year old filly 'Arctic Moon' which was to be her first Connemara pony. Since that time, she has bred and exhibited these ponies with spectacular success in England, and other countries to which they have been exported under the Chiltern prefix.

In a short time she was to join the panel of judges and colt inspectors for The English Connemara Pony Society, as a result of which she has acted regularly as a judge at the Premier English Shows, as well as receiving invitations to judge at The Royal Dublin Society and other shows in Ireland including Clifden.

She was editor of the Connemara Chronicle when it was first published in 1974 up to 1986. She has been a regular visitor to the Society's annual Connemara Pony Show at Clifden, and is well known to the majority of breeders in Connemara and, for that matter, all over the world. No one person was better qualified than Pat Lyne in 1984, to publish "Shrouded in Mist", the first authoritative study of the Connemara pony, in which she shows firstly how the breed evolved from its earliest days, and secondly, how the Irish Connemara Pony Society and its stud book were born. She follows the progress and the development of the Society and its stud Book up to 1963.

Six years later, in 1990 "Shrouded In Mist" was followed by the publication of "Out Of The Mist" which was to be "a further study of the Connemara pony throughout the world wherever he is bred and used", and to bring the story up to date from 1963 where the first book left off.

Now, hardly four years have elapsed since the publication of "Out Of The Mist" when the indefatigable Pat brings to press, this, her third book, "Reflections" which is not only a further updating of her chronicle of events in the progress and development of the breed, but what I would like to term a 'professional overview of the seventy year life of the Connemara Pony Breeders Society,' started by that small but brave band of men way back in 1923.

Breeders from all over the world, and those who love the Connemara pony, will remain deeply indebted to the author for her untiring efforts in passing on to her readers her deep knowledge and appreciation of the story of the Connemara pony, and of the people of the West of Ireland whom she knows and understands so well. For those who wish to know this story and keep a permanent reference to it, Pat Lyne's three books **will be** a must.

Long may she keep on writing.

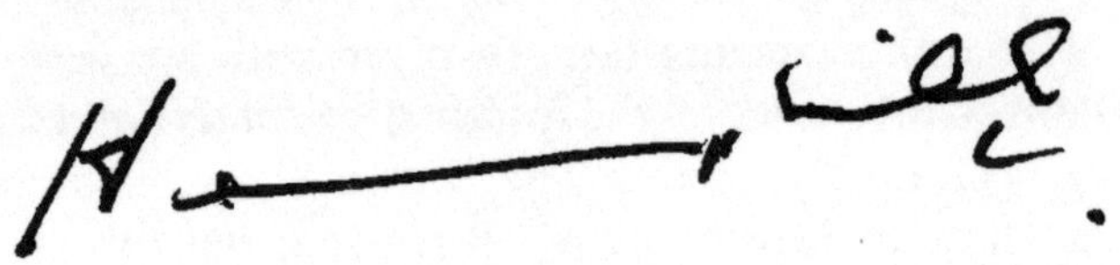

RAFORD. 1993.

CONTENTS

John McLoughlin and his mare MILFORD WREN (Sarsfield - Mervyn Wren) winner of her class at Clifden 1993 (photo Connacht Tribune)

CHAPTER ONE

MILESTONES

It is, of course, Arctic Moon I have to thank for my love of the Connemara pony and the Western Seaboard of Ireland. I met her by chance in 1964. I was on holiday touring Ireland with a friend. I remember our rental car was a brand new red mini which did a vast number of miles to the gallon. We set off down the East Coast staying a night here and a night there as the fancy took us. Arriving on the West Coast, for our last few days travel, we found ourselves in the main street of Oughterard. Rather than driving further into Connemara, with the limited time left to us, we decided to seek out the local pony breed.

An enquiry of the first person we met led us to Terry Miller at Ardnasillagh. She made us welcome, and we saw her two mares and foals. She encouraged us to visit the Geogeghans of Canrower, (Geogeghan being the third name in the space of an hour we could not get our tongues around) where, Terry assured us, we would have the opportunity of seeing a herd of ponies. We could hardly have wished for a better introduction to the Connemara pony than we received at Canrower. We were greeted by Bridge Boy's[1] head which shot out of a hole in the wall to give us a noisy welcome. While Paddy brought him out for our approval, his father Mark saddled a three year old gelding and set off into the distance, returning some twenty minutes later with a collection of ponies who walked peacefully ahead of him in our direction. There were old and young, some with foals, mostly light or dark grey and led by a plain rather worn looking bay mare! We were slightly bemused by the variety and trying to sort out who belonged to whom. Indeed we were embarrassed when we discovered it was assumed we were looking 'to buy' and not merely as interested tourists.

Mark returned the herd to the hillside and we were taken to see two ponies grazing in a small nearby enclosure. They were three year old fillies, one of them a golden dun whose name was Arctic Moon.[2] My eye, on which I relied to earn my living, told me this was a lovely pony. Her near perfect make and shape, her beautiful head and full eye, were under-pinned by a quiet charm. Almost instinctively I enquired her price. Unlike all the others we had seen she was not for sale.

We were taken inside to meet Mrs. Geogeghan and the family and given a welcome cup of tea with as much freshly baked Irish soda bread as we could eat. While Mark[3] sat silently in his high backed chair beside the turf fire, Paddy told us of Bridge Boy's success at Clifden show and of Silver Lining bred at Canrower, who was winning prizes in England. We were surprised to learn the plain bay

[1] O.O.M. 187-188 [2] O.O.M. 91-93 [3] S.I.M. 108-109

mare, Canrower Lass, was the dam of Arctic Moon, and full sister to Silver Lining.[4] Before saying our goodbyes I gave Paddy my business card and said "Let me know if you change your mind about the dun filly". As we left the youngest Geogeghan was playing happily on the hearth mat, cutting her teeth on my card. There was no reason to think I would ever hear from Canrower again. We spent our last night at the Shelbourne Hotel in Dublin, and it cost us all of £6. 18. 3d each.

I returned home to the usual busy round of horse shows, making and breaking young horses, pony club camp and the prospect of a busy last hunting season at Westmead. By early September the twenty loose boxes were full of fat hunters and I started on the slow grind of getting them fit. The happy memory of my Irish holiday and the small dun filly soon slipped from my mind.

In mid October, an envelope with an Irish stamp, dropped through my letter box. It was a letter from Paddy Geogeghan offering me the dun filly for £280.00. By this time the holiday spirit had worn off. There were, as usual, no spare pennies in the bank and I was on the brink of the hunting season. It would have been very easy to reject the offer. Instead, I joined the English Connemara Pony Society, received some good advice from Louise Barthorp and after a protracted haggle by post bought Arctic Moon for £260.00. She arrived at Bletchley station on a dark December afternoon four days before Christmas. As she stepped from her railway box she looked a very expensive purchase. She had grown a heavy winter coat, was smaller than I remembered and showed no sign of the foal she was said to be carrying.

We moved from Westmead to Vale Farm, Kimblewick, in the autumn of 1965. My mother gave up farming and I reduced my livery business by more than half. Arctic Moon and her filly foal moved with us. They were not a priority and took second or third place to my varied interests. Arctic Moon simply melted into my life and it was some years before she took centre stage.

In 1966 I placed my yearling filly Lune Star by Bridge Boy on the open market. There was no interest in her at £100.00 and so she remained with me. By 1968 there were two mares to send away to stud. This is when I began to toy with the idea of having my own stallion. For the second time chance played its hand.

For some years, George Mullan of Dublin was my trusted agent when buying horses in Ireland. Many a time we had scoured Co. Meath together, taking it in turns behind the wheel of his black beetle Volkswagon. George had the right contacts and a shrewd eye for a good young horse. He knew where my priorities lay and I did not take his August telephone call lightly. He was aware of my new interest in Connemara ponies and rang to say he had bought "one hell of a nice

[4] S.I.M. 227

Connemara pony stallion" I caught a plane to Dublin the next day but to George's dismay found myself unable to make an instant decision to buy Island Duke.[5] I felt I could overlook his slight imperfections and was very taken by his personality and what I recognised as native pony character, but I could not escape the fact that he was a "pig in a poke". Why had his Australian buyer left him behind in preference to another? He was six years old. Why did Mr. Killeen have no knowledge of progeny by him? He had to be inspected for the English Stud book and his pedigree was unpopular. There was little to encourage me to buy him except ...... my own intuition. I collected Dougal (as be became affectionately known) from Birkenhead docks in October 1968. I never for a moment regretted my decision to buy him. I registered the Chiltern prefix and thus crossed the Rubicon. I was now committed to breeding Connemara ponies.

These reflective milestones are not intended to be my life story, nor indeed that of the Chiltern stud or Arctic Moon. Rather, they are the signposts which led me to where I am now, writing these reflections.

In 1970 I was elected to the Council of the English Connemara Pony Society. It was also the year of the International Conference of the Connemara Pony Breeders Society in Galway. Following a visit in 1969 to Russia with a lay apostolic group, Poland was pencilled in my diary for a similar trip in 1970. An invitation to join my friends Honor and Pam, who planned to drive to Galway towing Honor's caravan called for a change of plan. It was an unforgettable week for the three of us, and what is more, we were still friends at the end of it! We parked the caravan in a small stonewalled enclosure over-looking Galway Bay, somewhere between Barna and Spiddal. The sun shone the whole week through, and the only thing we lacked was a bath, readily supplied by courtesy of John and Cecilie Williams at the Great Southern Hotel. It was there we registered on the evening of Wednesday, August 19th. The following day we made for the Clifden show ground.

I had no preconceived ideas of Clifden show but recall being overwhelmed at the number of ponies - four hundred and eighty were catalogued - who streamed in and, in some cases almost as quickly out, of the ring. At the same time I was charmed by the uniquely informal atmosphere and the rough and ready appearance of some of the ponies and their handlers. It was apparent they had walked in straight off the mountain, which we do not see in the Cliften ring today, and I have to confess to a certain nostalgia for this.

It is easy to forget I was very much a newcomer in 1970. I doubt I owned an Irish stud book and my knowledge of sires and blood lines was distinctly limited. None the less I relished the challenge of making judgements and sharing them

[5] O.O.M. 53-55

with my friends. When I missed a good one they found it, and vice versa. As I thumb the pages of a marked catalogue I realise many ponies in the ribbons that day became 'household' names as my interest in the breed grew.

It was customary in those far off times for the stallion class to open proceedings and with twenty-seven entries the ring was stretched to its very limits. It is always a class of intense speculation and interest, and I believe it sits more happily mid way through the day where it is now scheduled. By this time the show ground spills over with spectators and the fierce competition between the stallions, which never fails to thrill, can be appreciated by all. Their natural skills of showmanship ooze from every muscle and their handlers appear to despise a tight rein. It is a class never to be missed.

In 1970 the classy five year old Kimble[6] toppled the many times champion Bridge Boy, who had to be content with third place. The 'confined' stallion class, sadly no longer part of the programme, also fell to Kimble and I notice the three year old Killyreagh Kim[7] was fifth of seven. Thirty-eight yearling colts were listed and the winner, a son of Rebel Wind[8] and Breath of Spring, must surely have been Dangan Dun. The Fugitive had no difficulty in taking the two year old class from Roundstone Oscar. The entry of eighty-four yearling fillies provided a show on their own. John Brennan set his heart on winning this class and, discerning breeder that he was, took 1st, 2nd and 4th place with his three entries. At this time, of course, the fillies were unregistered and unnamed, but they were Fort Viola, Fort Holly and Celebration Cream. Breath of Spring won the five and six year old mare class for Josie Conroy and Cashel Kate[9] the seven years and over with foal for Mr. Donnelly. She was followed home by Jack Bolger with Finola of Leam.[10] There were only twelve entries for the barren mare class (no longer judged separately) which was won by Atlantic Mist. She went on to take the supreme with Kimble standing reserve to her. I notice the following year Kimble was supreme with Fort Holly in reserve. How well the form stood up. From a galaxy of foals twenty year old Cregg Lassie[11] provided the winner of the Camlin Ciro Challenge Cup, a filly by Bridge Boy. We had been treated to a display of the very best of the young and the old in Connemara in 1970. We returned to our caravan bunks dreaming of all we had seen that day.

On the following morning the Conference was opened in Galway University by the Minister of Agriculture, after which Lord Killanin presided over a tightly packed two-day programme attended by many overseas breeders. Papers were given by experts on Equine Disease and Nutrition and Training the Pony. A ten minute report was invited from each Society present, and an open discussion followed. The second day concluded with a Parade of stallions at Tulira and a reception at Dun Guaire Castle. As I shake my kaleidoscope of memories two or three

[6] O.O.M. 221-222
[7] O.O.M. 271-272
[8] O.O.M. 271-272
[9] S.I.M. 215
[10] O.O.M. 25-26
[11] S.I.M. 229

rise to the surface; Sean Keane's short history of Connacht; Lal Berridge's report on the forty-seven year life of the C.P.B. Society; the Tulira Parade, and a private visit to Grey Dawn Lassie. Each of them sowed a seed in my being which germinated slowly over the next few years.

Tulira Castle provided a fitting back-cloth to what was surely an historic occasion, a parade of the founding fathers of the breed, those who had sired many of the best we had seen at Clifden show. The old ponies walked sedately round the prepared area in order of seniority. How regal they looked wearing their weather beaten, elderly clothes. Twenty-four year old Carna Bobby[12] was followed by his son, Ballydonagh Rob: twenty-two year old Carna Dun[13] by his seventeen year old son Doon Paddy: twenty-one year old MacDara[14] and his son Jolly Roger: sixteen year old Clonkeehan Auratum,[15] and thirteen year old Thunder.[16] The sight of these grand old ponies brought a lump to my throat and this parade gave me the first inkling of what breeding Connemara ponies was all about.

One eventide we visited Grey Dawn Lassie, dam of Island Duke, and her owner Patrick Mulkerrins. Patrick, a stallion custodian since 1958, had Dun Aengus[17] in his care and Lassie had a foal at foot by him when we visited. We took the Moycullen to Spiddal road and followed a track to the left which, we were assured, would lead us to Knockerasser. As we bumped our way along its uneven surface we wondered where we would turn if it led nowhere, which seemed increasingly likely. We appeared to be driving into the sky, with nothing but bog on either side of us. As the track turned and dipped we were encouraged by a gentle curl of smoke. Surely it came from a turf fire? Around the next bend sat two homesteads, the old and derelict beside the modern bungalow. We found Patrick and his Mother in the latter.

Lassie was in her twentieth year and as she trotted proudly against the sky line with her foal beside her and the sun going down behind her, I was aware of a matriarchal beauty which disguised her age and the hardships life had inflicted on her. She said "Look at me and my foal - and catch me if you can." I was very moved to be standing on "Dougals" birth turf. Ten years later when I unravelled Inchagoill Laddies[18] character I realised it was he, her sire, who bequeathed Lassie with an indefinable air of belonging exactly where she was. It was she, Grey Dawn Lassie who inspired me to research the role of the females and their families for Shrouded in Mist. How was it I did not include her? A fear of being charged with nepotism maybe?

When we came to leave our caravan site I felt sad but saturated with all that the last days had provided. I sat on the back seat of the car and opted out of conversation and suggestions of further visits. I felt I had no room for more. I had lost

[12] S.I.M. 159
[13] S.I.M. 189-190
[14] S.I.M. 171-172
[15] S.I.M. 196-199
[16] S.I.M. 180
[17] O.O.M. 36-38 + 273-274
[18] S.I.M. 151-152

my heart to Connemara, its people, its history and it goes without saying, its ponies. I had much to dream about.

In 1971 the Chiltern Stud, Jo and I moved from Buckinghamshire to Redgates in Herefordshire. It was a time of great change. We were moving from five acres to thirty-five, and breeding Connemara ponies became a reality. At the same time I was impatient to return to Connemara and to share all it had to offer with Jo. And so later the same year we planned our holiday to coincide with Maam Cross Fair. We stayed in a B & B at Maam Bridge and spent our days exploring Connemara. We found we were equally happy in the shadow of Maam Turk or the Twelve Bens or among the fuschia hedges of Kylemore or walking the golden strands of Mannin, Roundstone or Omey. Each area had its own special features.

Maam Fair is as traditional as Clifden Show and is held on the last Tuesday of October at a cross roads which could well have been designed for this very purpose. The width of the road and the open space either side of it afford unfettered freedom for the loose foal, the unhandled yearling on the long rope, and the ram, attached by his horns, to the nearest post. Peacockes forecourt used to be the sought-after 'pitch' close to the bar and the centre of things. For the buyer, the chance of a better look at a pony can be had down at the cross or beyond.

The October day can dawn bright and clear or wet and damp and sometimes foggy and frosty. When the sun shines the gorse glistens, the Twelve Bens smile and the ponies look twice their every day value. In the wet, all backs are hunched and turned to the prevailing wind. Manes and tails hang lank and dank, the Bens glower and it takes a few extra jars for business to get under way. On a frosty morning, the breath from the ponies nostrils and the steam from their flanks rise in unison with the smoke from the pipes and cigarettes of their handlers as they hit the frosty air together. Business is brisker, the Bens are shrouded in fog and the jars flow freely. It is the day for a crack and a deal and there is plenty of good will and enjoyment among those who are there to sell or to buy or merely to look on.

In 1971 my cheque book was in my pocket and I was there to buy. It was a cheerless, nondescript sort of a day but this did nothing to dampen our spirits. We met seventeen year old Honey Belle with her owner John Walsh Pat. They had walked in from Rosmuc in order to sell her colt foal. Across her shoulders, Honey Belle carried a sack of oats which John Pat offered her before their long, lonely journey home. Further up the road we met twenty-six year old Sliabh Na mBan 2nd[19] who had travelled in from Ardmore without a foal and who was unlikely to return home that night. The fair has always been the pony-man's sale yard where he expects to wean and sell his foal or to see his old mare to her final destination. These partings are made without sentiment. There is no room for such in the hard

[19] S.I.M. 211

graft of life in the West. By ten o'clock we were in the bar sharing a drink with Paddy Moffit and trying to do business with him. We made friends with James Healy of Salt Hill who bought us our second drink of the morning. We did not meet Don McBride of Westport until the end of the day. He appeared to be on his way home not having had a deal. "Would you be interested in this little mare?" we were asked. I noticed she was shod and had rather a plain head but she trotted up the road freely enough and looked keen and intelligent. Don was eager to tell me she was by Errigal Prince but was a mite embarrassed when he admitted her dam was named Florence Nightingale. Silver Lass was five years old and said to be in foal to Little Joe.[20] This conveyed little to me. I found the asking price was easy to negotiate so late in the day. Silver Lass returned to Westport with my initials on her rump and travelled to England two weeks later with Tully Blue.[21]

I have returned to the Fair several times since 1971 but not to buy ponies. The temptation to do so is always there and I confess my pulse does not race quite so quickly when I am a mere onlooker. In the 1990s I hear of Marts conducting sales and of catalogues being prepared. I fear the Irishman's traditional way of doing business on a one to one basis is fast being eroded and that Maam Cross, which, together with Ballinasloe is one of the last remaining Irish fairs, will lose its individual place in the life of Connemara and become just another sale yard.

In 1973 I was invited to join the E.C.P.S. Judges Panel, a privilege I was happy to accept. I had worked my way through the tough school of the British Show Hack and Cob Association probationer scheme and had been judging show hacks and riding horses for some years. At around the same time the future of the Society's Quarterly Newsletter was in question and the Council decided to replace it with a yearly Journal. I was asked to produce the magazine and was given £50.00 towards the cost of the first edition printed in 1974. And that is how the Connemara Chronicle was born. I had no qualifications for the task of Editor but it presented me with an entirely new challenge which I enjoyed. It stimulated my interest in many wider issues of the Breed and was certainly the source of my inspiration for Shrouded in Mist. It was like losing a good friend when I resigned as Editor, after twelve years at the helm.

The third challenge of 1973 was to find a colt to use on my Island Duke fillies who would be ready for stud the following year. We made it an excuse to visit Clifden Show to find one. At the same time we realised it was unlikely we could afford a colt in the top end of the yearling colt class. In reality there was not a single colt in the class who fulfilled all our needs and as the week went by we became aware we would be lucky if such a colt even existed. After the show our enquiries led us to the Ballyconneely district where we met Paddy King, Michael Conroy, Thomas McHugh and others. Each one became intimately concerned

[20] S.I.M. 190 [21] O.O.M. 23

with our desire to buy a yearling colt and pointed us in this direction and that, and even accompanied us to make sure we did not get lost. We made lasting friendships on this visit. I learnt my way around Bunowen, Lehid, Truska and Doonloughan, some of the delightfully hidden areas of Connemara and Ballyconneely, always a centre of pony breeding. To see the herd of ponies grazing the Truska beach presented a new aspect of Connemara pony life, which contrasted sharply with the bog side and the mountain top.

We left Ballyconneely with a strong feeling we would settle for the Killyreagh Kim colt but wished to see others before making the big decision. So much depended on it and I doubt there is anything more difficult to buy with confidence than a Connemara yearling colt.

We moved on to Cashel and at midday on Sunday we knocked on Jack Bolger's back door to enquire whether he had any ponies for sale. When he replied 'Yes but they were a short step away' we took him at his word. His grand-daughter was our guide and the short step was a good hour's tramp across the bog through a heavy, damp mist. When we eventually found the herd we startled them. Their heads and their tails went into the air and as they galloped away from us I had a fleeting glimpse of a grey stallion leading a motley selection of mares. Our young guide did not know who any of them were. One mare caught my attention by the effortless, free way she moved across the rough terrain. She was easy to identify for she was bay with a white blaze. By the time we returned to Cashel we were tired, wet and hungry. Jack told us the bay mare was Common Market by Rebel Wind.

The memory of our final visit on our return to Galway is lost in a haze of potcheen, a drink to be avoided when you have a cold, empty stomach! The following day we contacted Mr. Healy in Salt Hill and were invited to see his yearling colt by Finney Master.[22] Imagine our surprise when we discovered his dam was Celebration Cream, grazing in the field next door.

When we visited Paddy Moffit in 1971 his little mare, Happy Rebel, and her two cross bred gelding sons took my eye. She was carrying a foal to Kimble. I asked Paddy to retain it for me. A phone call confirmed he had kept his word and the yearling filly, Tubber Posy, together with one of the geldings travelled to England three weeks later with Truska Pimpernel,[23] Ice Cream and Common Market. What a consignment and how I agonised over the risks I was taking! I had been unable to get close to any of them. I could only trust in what I had seen, or knew of, their respective sires and dams and yet again, on my own intuition and also on a gradual, growing awareness of the qualities of the breed. All the ponies fulfilled my hopes for them but it was, of course, a long, long wait for their fruits to reveal themselves.

[22] O.O.M. 224-226 [23] O.O.M. 29

In 1976 an unexpected bonus came my way. A free ride to Connemara. I was asked to accompany John Meade to meet our parent body in Galway to discuss the harmonisation of stud books. A directive from Brussels instructed all breeds to heed one paragraph in the Treaty of Rome "It is illegal for members to operate stud books that are not common to all other member countries of any breed." Our two meetings with Sean Keane, John Killeen and their Council are now old history. And the ICCPS[24] was not even a whisper in the wind!

I travelled to Galway with John and Phyllis where I collected an Avis Car. It had bald tyres, no spare and spluttered ominously but carried me without mishap for many Connemara miles. On our first day we had lunch at Tulira, after which Anne Hemphill took me to visit Clonkeehan Auratum. He was in his twenty-second and last year of his life. He was wearing, what appeared to be, a strange alternative to hobbles. A rope halter was attached to a surcingle type belly rope, acting like a tight standing martingale. If it was meant to restrict his movements it was not very effective. As I endeavoured to take photos of him he moved very fast in the opposite direction showing a proud defiance. I treasure that last glimpse of a brave old pony. Our call on Clonjoy[25] is not a clear memory but a tour of the Tulira ponies revealed how much Anne had achieved in fourteen years of breeding Connemara ponies.

I was anxious to make good use of these few precious days and as always turned to my friends Steph and Garnet for help. I spent time at both Errislannan and Ross, and shared my ideas for a study of the breed, and maybe a book to follow. My instinct was to meet and talk with as many of the old breeders as possible and to build up a general picture from them.

I remember driving out to Carna for the first time and realising how much of Connemara I had yet to discover. Each new area I visited revealed its own scenic charm which cannot be experienced if you do not move beyond a ten mile radius of Clifden. The scape of the inland road from Glinsk to Carna is a rolling mass of rock and stone with no shelter for man or beast. The coastline is softer and the Kilkerrin harbour and its off shore islands provide a beauty of their very own, adorned by the local fishing boats as they set out to sea in a failing light. I find my notes for 1976 are sketchy but those early, personal visits to Callowfeenish, Cregg, Shanbolard, Screebe, and return ones to Bunowen, Cashel and Canrower remain an indelible memory. It was the friends I made on this occasion who enabled me to fashion the fabric of Shrouded in Mist. Steph and I spent a day at Maam Fair and were delighted at the sight of John and Phyllis Meade falling to the charms of a filly by Ceilog, whom they succeeded in buying before the day was out. Two nights at Ross were spent pouring over old stud books, records and photographs, while Garnet's knitting needles clacked merrily away in the background.

[24] International Committee Connemara Pony Societies

[25] S.I.M. 198 & O.O.M. Index

We had a noisy, uncomfortable night-boat return journey to Liverpool. Somehow it did not matter. I was on my way home with a determination to return as soon as possible to unravel the history of the Connemara pony. My determination may have been dented had I realised how long and stony the road of research would prove. When my courage failed me I would say "Fire Up" just as Cannon Balls[26] jockey urged him those many years ago.

The resolution to return was very strong. An advert in Horse and Hound "Cottage to rent in Connemara" was all I required to spur me on. We made a booking for three weeks, September 10th - October 1st 1978 and remained in Ireland for a further week. I cannot imagine how we were able to leave Redgates and the ponies for a whole month. Tansy, our sheepdog, went with us.

Our first impressions of the cottage were not encouraging. After a long search we found the front door key buried in leaves in the gutter over the front door. Inside we found the cottage dirty, dark and poorly furnished. After a while we learnt which chairs were safe to sit on and which beds were less uncomfortable than others! We coped cheerfully with the lack of water, fetching it daily from a hand pump in Letterfrack, two miles up the road. Eventually kind neighbours refilled the empty rain water tanks for us. We learnt to light and enjoy the comfort of a turf fire. Jo got the best of an indifferent cooker. Never for a moment were we anything but happy. I am so glad Jo kept a daily diary. Here are just a few snippets from it.

Sept. 15th. Visited Graham Tulloch. Busy making silage. Showed us his two year old colt Loubine Larry (Loobeen!) and mares due to leave for Brittany. Water tank empty p.m. Leg fell off chair. Handle off kettle. Picked blackberries.

Sept. 17th. Met Mikey King. Took him to Claddaghduff bar and played darts. Saw "Sarsfield" grazing in hobbles. Liked him.

Sept. 19th. Found lovely Dogs Bay. Met James de Courcy and family and ponies. Bought chops for Sunday lunch from Tommy McDonagh. Saw his valuable old photographs of ponies.

Sept. 21st. Visited John Daly, Lough Mask House. Had a great day. Called on John Luskin and Milford Hurricane.[27]

Sept. 23rd. Drove to Cashel through Inagh Valley. Spent time with Jack Bolger and drove on to Callowfeenish and the Mulkerrins. A real feast.

Sept. 24th. Visited Oughterard. Mark Geogeghan took us for a drive in a sidecar by Lock Corrib. Had a meal with Jim and Nora Walsh, Cregg. Kept evening appointment with Graham. Had his grandmother's photographs, a warm fire and whiskey all ready for us. Stayed till 11.30 p.m. great day.

Sept. 25th. Mass at Letterfrack. Lunch at Errislannan. Met John Folan.

[26] S.I.M. 147-150 [27] O.O.M. 266-267

Sept. 27th. Visited Cleggan coastguard.[28] Followed the sky road to find Muffy. Tea at Glasthule with Kitty Russell and Jardine.
Oct. 1st. Washed our linen. Cleaned house. Were sorry to leave "the happy hovel" Farewell picnic lunch on coral strand. Great to be welcomed at Maam Bridge by Nora Keane p.m.
Oct. 2nd. Left for Rosmuc after Mass. Met Mattie Mannion. Down to Walsh's Bar where we watched the Consecration of Archbishop O'Fiaich on T.V. sipping a gin and tonic.
Oct. 3rd. Kept appointment at Connacht Tribune and with Mr. Killeen. Had first look at Society's stallion records in battered file. Supper at Cregg. Left with a bag of 'Jim grown' potatoes.
Oct. 4th. Left Maam 9.30 a.m. very sad. Called to say goodbye to Mark. Drove to Spiddal via Costello, Glenicmurrin and Bealadangan meeting Bartley Naughton and John Connolly. Went to Carraroe to see Rebel Wind. No luck. He was away on the mountain. Michael Clancy and Jim Lee not at home. Spent the night in a B & B at Loughrea.
Oct. 5th. On the road early for Abbeyleix. Had such a happy visit. Left for Slane p.m. Made it by 7.00 p.m. as arranged.
Oct. 6th. I was rather overawed by Miss L. N.* and Tansy by her pack of dogs and hunting whip control. Made welcome and comfortable. Met Clonkeehan Nimbus and Yeoman. Went to Dublin to photocopy Susan's valuable records. Dan L. O'K.* came to supper. Quite a day.
Oct. 7th. Left for Enniskerry and Lal.* Arrived for lunch. Stayed on and invited back for supper. Such a happy day. B & B in Bray.
Oct. 8th. Our last day in Dublin, Pat had purse stolen in Easons. Went to see Lisa Minnelli in 'New York, New York'. Spent the evening in Police Station reporting theft. Waste of time.
Oct. 9th. Spent a quiet day. Picnic in Greystones. Caught the night boat to Holyhead.

I returned home with copious notes, poorly recorded tape conversations, some precious photos and "fired up" with enthusiasm to make a start on Chapter One. The enthusiasm was soon dissipated in the demands of daily life, stud, farm, parish, E.C.P.S. Council, Chronicle and Jo's poor health. And so all I gleaned was stored away in a bottom drawer, in my memory and in my heart and once again a book was no more than a dream. Jo and Connemara friends in England never ceased to encourage me. In 1978 we planned another autumn visit to Connemara in a rented cottage and this time we were oh so lucky. The owners of Station House, Recess had converted the old waiting room (or was it shunting shed?) into a comfortable flat for family visitors. Occasionally it was 'to let' and we were lucky

[28] S.I.M. 113

* Francais Lee Norman, Dan Llanigan O'Keefe and Lal Berridge

enough to spend two happy weeks there. On our way to Recess we made calls in South Connemara to Oranmore, Kilcolgan, Ganty and Ardrachan. To our delight we found twenty year old Clonjoy happily grazing rough pasture which was overrun with prickly sloe bushes around which he played hide-and-seek until he decided it was safe to talk to us. We made another fruitful visit to Mr. Killeens office where we were now accepted with plenty of goodwill and trust.

Top of my priorities in 1978 was a visit to Frank O'Malley in Rosmuc. I was impatient to see a copy of his Father's letters.[29] Because no one had expressed an interest in them for so long, Frank was uncertain if he still had a copy. What a relief when he discovered one battered booklet in his attic trunk. I took it to Galway and had a photocopy made. It was not very sharp and the photos of poor quality. It was a treasure of untold wealth to me and later, when Frank knew me better, he lent me the original. I'm sorry more people have not shown interest in the reprint we undertook in 1986.

We spent our time in much the same way as in the two previous years, travelling the Connemara roads, meeting as many pony breeders as possible, filling in cracks, finding new ones and never failing to unearth a small piece of history. It was a painstaking but rewarding task in many respects. We spent a day in Co Mayo and another at Maam Fair, always a rare opportunity to meet anyone and everyone. On this occasion it included a film crew, whose most recent assignment had been the Kentucky yearling sales. What a contrast! On our last evening at Recess there was a knock at the front door. It was Padraig Hynes from Canal Stage with an invitation to see his ponies. We called the following morning and met his mother and his mare Rambling Home. Padraig has been a steadfast and valued friend ever since.

I realise, with hindsight, the O'Malley letters were a turning point in my research. No longer was I content for Chapter One to open in 1923 with the foundation of the Pony Breeders Society. This extended the time spent on research but I hope it was worth it. Before we left Station House I had completed a rough draft for what became Chapter Seven "Meet the People, Know the Place". I had a sense of getting somewhere at last.

As always we drove out of Galway with sadness in our hearts. To try and cheer us up Jo suggested we find a title for the book. Jo did most of the finding and wrote down twelve suggestions. I did the driving and the choosing and Shrouded in Mist won the day.

We returned to Redgates and a similar situation to the year before. I had embarked on a Farm Improvement Scheme with tempting EEC subsidies but inevitably it increased the work load. I 'bartered' a cremello filly for twelve

[29] S.I.M. Ch. 6 95-103

breeding ewes, (I knew nothing about sheep). We prepared a home bred Aberdeen Angus cross steer for Hay-on-Wye Christmas Fatstock Show (he was champion). We expected our usual quota of foals, visiting mares and show engagements in 1979. It remained a relentless struggle to earn a living from thirty-five acres.

In August I celebrated my 50th birthday and in October I returned to Station House, this time on my own, for a little over two weeks. I settled to the task of sorting the material I had gathered in the last years, building on it and making a positive effort to write. It was a fruitful visit and I could have worked away happily for two months, never mind two weeks. Each line I wrote revealed gaps in my research and I was tempted to jump in the car and drive to wherever necessary to find the missing link. I found the research more to my liking than the writing, which at times did not come easily. Each visit brought me closer to an understanding of the people and their life style, and the ponies place in that life style, both in the past and in the present.

During this visit I discovered a Redemptorist Priest giving a Mission in the small wayside church at Ballinafad. I shared daily Mass and the Mission theme with the Parishioners each evening. It was a welcome source of support at the end of a day's labour with the pen. The mission closed with a Harvest Festival. The fruits of the earth were arranged around the simple altar with the men of the Parish, in their Sunday best, carrying the tools of their trade. The spade to dig the potatoes, the slane to cut the turf and the fork to toss the hay. They stood among the fruits as thanksgiving was offered. This little church of St. Joseph, built in 1835 and enlarged in 1911, must have seen many years when there was no harvest for which to offer thanks.

My interest in the influence of the Irish Draught on the Connemara breed led me to Michael Duane of Kilreekil, Loughrea, Secretary of the Galway branch of the I.D.H.S.[30] I always received a large slice of home made apple pie at Kilreekil, and it became a regular watering hole. His Champion mare, Pink Carnation, invariably had a good foal to show me. Michael provided another 'cog' in my wheel of research and I was sad to read his obituary in the Connacht Tribune in 1987. He was remembered as a 'tasty farmer, a skilled ploughman and a front rank judge of horses.'

1980 is a year, which in many ways, I would rather forget. How can I say this when I had three splendid weeks in Australia in April and judged the Connemaras in Dublin in August?

For ten years we had enjoyed and suffered the usual swings and roundabouts of pony breeding. The mares were regular breeders, the stallions proved their worth, we sold young stock abroad and were in the ribbons at all the major shows.

[30] Irish Draught Horse Society

We worked hard but were well rewarded and enjoyed - almost - every moment of it.

In 1980 from five foals we had four colts and a blue eyed cream filly. A two year old colt (unrelated to our mares) proved infertile. Once we realised we were wasting time we sent two mares away to stud. Neither held to the stallion they visited. Dear, reliable Lune Star gave up breeding. Island Duke died on October 18th. I resigned from the Council of the ECPS in July.

I travelled to Dublin the day before I was due to Judge and stayed with George Mullan's widow in Sutton, Co. Dublin. It was a happy reunion for us. Show day was the wettest and coldest August day you could imagine. Michael Clancy was my co-judge and neither of us could sign the Judges book for our numb fingers. None the less, it was a rare privilege to be Judging at Dublin and the hospitality of the R.D.S.[31] was second to none. I was invited to join the Brooks family for dinner that night, a grand way to end the day. I received a letter from a friend in England after the show. "I saw you in the distance in Dublin" she wrote, "in a sea of mud and a sea of Connemaras". And that was how it was.

I arranged to work in the R.D.S. Library the following day. It was there, thanks to head librarian Mr. Eager, I learned what the Congested Districts Board[32] was all about. It made very interesting and valuable reading. I was free on Sunday and drove out to Co. Longford to meet Tully Grey[33] and his owners. He was an old pony by this time but I am pleased I had the opportunity to see him before it was too late. It was hard to be so near and yet so far from Connemara for these four days. Time did not allow a visit but my stay in Dublin was far from wasted. Another chapter for Shrouded in Mist had taken root.

In 1980 Ros Harris, a comparative newcomer to the English Society, arranged a lunch party at her home for E.C.P.S. members living in the South East. I was invited to give a talk. "The Connemara in his native land" was the chosen theme. Ros had great enthusiasm for the breed and was a loyal supporter of The Chronicle. She was also well aware of the frustrations I suffered at the stop/go regime under which I was working on Shrouded in Mist.

There was little improvement in the fortunes of the Chiltern Stud in 1981 apart from Dougal's last gift to us - two fillies born to Arctic Moon and Common Market. In May I met with an accident handling a yearling colt and injured an ankle. I was in plaster and on crutches for some weeks but the full extent of the damage done was not discovered until an operation in 1985. For most of 1981 it made life painful and difficult.

The same year Bent Nielsen staged a second Connemara Festival in Denmark. I entered a two year old colt Chiltern Celtic Dancer. It was a costly exercise to transport him to Roskilde. I believed he was a good pony and took a sizeable

[31] Royal Dublin Society [32] S.I.M. Ch. 2. [33] S.I.M. 179

gamble on the chance of selling him there. He won his class and was Reserve Junior Champion but as anyone who was present will remember, it rained heavily and remorselessly for the two day show. Any hope I had of selling him was quite literally washed out. He came home, was subsequently gelded, and I was more than one thousand pounds worse off!

By this time I was grateful to have other things to take my mind off our difficulties. Ros Harris conceived the idea of raising sponsorship to enable me to take a three month sabbatical in Connemara in order that I might work on Shrouded in Mist without interruption. Many members of the Society generously supported her initiative and I shall always be more than grateful to Ros (and Ann Wilson who helped from her Scottish base) for all they did to make the dream a reality. I decided to abandon the Farm Improvement Scheme, the completion of which was anyway out of the question with an injured ankle. On September 6th we had a farm sale at Redgates and sold everything, bar the kitchen sink, the ponies and Celandine the house cow. On October 1st I left for Ireland. Don and Steph Brooks generously offered me the use of Errislannan's Gate Lodge for three months. Jo travelled with me to Connemara to see me settled in. And Tansy also, who was my faithful companion, and without whose company I would have been so much the poorer.

I drove Jo back to Dublin and spent three nights in Bray to keep some important appointments. I met Donal Kenny in Wynns Hotel, Lord Killanin in his Landsdowne Road, Dublin home and Tadgh O'Sullivan in his St. Stephens Green Office of Foreign Affairs. I had introductions to the Keeper of the Books at Trinity College and the Kildare Street Library. To read and handle old books became a real pleasure. Research was compulsive and I realised I had to 'let go' of it sometime soon.

On my return to the Lodge I settled down to work and my diary tells me how much I achieved each day. It was a period of contrasts, days of satisfaction, and of total isolation from the rest of the world; days of frustration at my lack of progress and of loneliness; days of despair when I felt the task I had undertaken was beyond me and how would I tell my supporters?

On the plus side I slipped very easily into the culture and life of Connemara. No longer was I 'just' a visitor but a resident and I felt "at home". I joined the Storytellers Club in Clifden and shared excursions to the ruins of Clifden Castle and the like. I was invited to give a talk to the sixth form boys in Clifden's Arts week on the history of the Connemara pony. They were very shy of me but one boy introduced himself as grandson of Stephen Walsh, owner of Silver Pearl.[34] I joined those who lived on the Errislannan Peninsular for a Station Mass at Mrs. King's post office on the Point. The counter was transformed into an altar and the

[34] S.I.M. 175-176

congregation filled the three downstairs rooms. The welcome I received from my neighbours when sharing this celebration warmed my heart. Mrs. King's hospitality, which followed Mass, was an extension of this traditional act of worship, which moves from one townland to another in certain seasons and for certain occasions, such as the blessing of a new home. For my relaxation I relied on the Irish Times and RTE, with its rich supply of plays and music. What a relief not to have a T.V. .... but I could not manage without my whiskey night-cap! Thomas Merton's "Seven Storey Mountain" provided my spiritual reading and was a very good companion in my hermit days.

Most of the time The Manor was unoccupied. When Siobhan and her friend were there they often shared a meal with me and I was glad of their company. Steph's arrival from London on two occasions was like a breath of fresh air. She came armed with interesting books and loads of encouragement. I was able to spend a day at both the Maam and Clifden Fairs. Twice I enjoyed warm hospitality at Bridge House Hotel with Michael Clancy while on visits to Galway. I was always assured of a welcome and a meal with the Hynes at Canal Stage or the Walshs at Cregg. I set myself a tight writing programme which was probably a mistake. I suffered various aches and pains and poor nights as a result. It was the coldest October in Dublin since 1896 and it did not feel very warm on the West coast either. I ate poorly at irregular hours but Tansy ensured I took some regular exercise each day. I was due home on December 21st with a lunch appointment at the British Embassy, with an old fox hunting friend, (currently English Ambassador), en route. In the event the onset of bad weather encouraged me to return earlier. My diary reads:

Dec. 11th. Woke up early. Very severe frost; water and gas frozen. Roads like ice. Did not dare take the car out.

Dec. 12th. Rang Jo. Her helper has left. Weather bad her end. Worked long and hard on mare studies. They are very absorbing and help remove my anxieties about home

Dec. 13th. Everything is running again. Roads are clear. Decided to pack up and leave.

My journey home was a nightmare. Once beyond Ballinahinch the roads deteriorated. It was Sunday and the whole of Ireland under a blanket of snow. I was just about the only car on the road between Galway and Dublin and broke down twice when the car overheated. I arrived in Bray at 9.30 p.m. only to find severe storms had broken the sea wall and access to my usual hotel could only be gained through the back door. No gas or heat in the hotel but Mrs. O'Connor managed to fix me up with a hot water bottle. I left without breakfast the following morning to catch the early boat and after a slippery drive to Dun Laoghaire was the last car to board.

When I drove off the Island of Anglesey road conditions became as bad or worse than in Ireland. All the Welsh mountain roads were blocked by snow. I travelled the longer coast road and arrived at Redgates in the dusk to find Jo at the end of her tether. She had managed, on her own, to get all the stock in but while doing so the back door had blown open and the house was full of snow. Altogether an inglorious end to my three month sabbatical.

It took some time to settle back into normal life at Redgates. Three parts, or a little more, of Shrouded in Mist was in rough draft. The farm and stud work load were reduced and time for work on the book easier to find. And still I did not complete the task. In August 1982 I returned to Connemara for Clifden show as Garnet's chauffeur and companion. It was a golden opportunity to find answers to newly revealed gaps in my research and to rekindle my resolve to complete the book. A whole year later on September 23rd 1983, I took the final text to Orphans Press. The index and all the art work remained to be done.

1982 and 1983 were unsettled years. We let twenty of our thirty five acres for grass keep. I became involved with other things. To help the falling income I took a part time job as a home help. I trained to speak on local radio and gave a series of "Thoughts for the Day" (unpaid!). I undertook to teach a Chinese mother English (unpaid!). I became more committed to giving regular help to the Carmelite Sisters at Presteigne. I took responsibility for their four acres, drove sick sisters to hospital and made myself available to help when they needed me. I began a four year period of formation as a member of the Third Order of Carmel. Jo and I considered the idea of moving to Presteigne and buying a property with less land than we had at Redgates. Nothing suitable came on the market. After a year's training in Cheltenham Jo was now a full time Health Visitor. By March 1984 Shrouded in Mist was ready for proof reading and that was an horrendous task. I was making heavy weather of it on my own and was kindly invited to stay at Carmel so that Sister Clare - who helped by typing the copy - could assist me and complete the work. During this week I saw the problems the Community were experiencing with their 'extern' help. On the feast of St. Patrick, March 17th I made the commitment to sell Redgates, and move to Carmel sometime in the autumn. It was a momentous decision. The months of preparing to sell Redgates and reduce the stock were unreal. The prospect of no longer sharing life with Jo and finding somewhere for her to live was difficult to come to terms with for both of us.

I set the E.C.P.S.[35] Breed Show at Stoneleigh in 1984 as my target for the publication of Shrouded in Mist. This was only achieved by a whisker. In August I packed the car with four hundred copies and drove to Ireland. The book received a chilly welcome in Dublin. The Irish Horse Board would not display or sell it on their stand as I had anticipated. Easons took only five copies. As I moved West the

[35] English Connemara Pony Society

situation improved. An evening launch party at O'Gorman's bookshop in Galway was a happy occasion. Kenny's also welcomed the book and I drove out to Clifden with a lighter heart. Clifden show managed to provide one of its warm, sunny days and I sat beside the Commentator selling books all day long. This was the day I had dreamt of for seven long years and remains a lasting, happy memory. I returned to Connemara in 1985 to judge at Clifden, but not again until the spring of 1989. In the years between Carmelites took precedence over Connemaras. Shrouded in Mist received encouraging reviews and remains in steady demand as stocks run low.

The move to Presteigne was made in September 1984. Arctic Moon and Chiltern Gazelle[36] went with me. All the other ponies were found temporary or permanent homes. Dragonfly moved to Kington with Jo and has lived happily at Woodbrook ever since.

My four years at Carmel were not easy, though as time passed I 'grew' into the way of life it provided and found a degree of contentment. In April 1988 it was decided the Community would disperse and the Carmel close. While it was a shock it was not entirely unexpected. The following twelve months were a painful time taking Sisters to their new Carmel homes and packing up at Presteigne. I moved down the road to Combe Cottage by Christmas 1988 and was grateful to have somewhere to move to. By 1987 I felt an urge to write a sequel to Shrouded in Mist and ideas for this began to take shape.

When the last of the Sisters left Presteigne, towards the end of 1989, I was quite bereft. Out of the Mist provided a welcome distraction without which I could easily have lapsed into depression. Sister Clare (now living an eremetical life in South Wales) continued to be my main stay and Out of the Mist came to fruition in February 1991. My Reflections complete the trilogy and will, I hope, be welcome.

[36] O.O.M. 92-93

CHAPTER TWO

CAMEO PORTRAITS

When I roll back the carpet of the seventy year life of the Connemara Pony Breeders Society I find many colourful characters within its weave. They are people who should not be forgotten in the passage of time but rather remembered for their enthusiasm for the Connemara pony and their part in his progress. Many of them are no longer with us. Others, happily, are, and I hope a visit to the portrait gallery I have assembled will stir happy old memories for some of us and inspire others to follow the example they set.

The difficulty was to find a starting and a finishing point and to judge how selective to be in my choice of portraits. To start in and return to Connemara seemed the best solution. Two names run a strong thread through the weave and it may appear odd that they, two ladies, should be hung first in my gallery. Odd, because, traditionally, the pony is the preserve of the man rather than the woman in Connemara although we find the reverse is the case as we travel outside the region. Garnet Irwin and Steph Brooks have put down roots, each of a different nature, in Co Galway. Lets look behind their portraits and discover how and why.

CONTRIBUTORS GALLERY

Garnet Irwin was an important member of the "Connemara racing set",[37] a unique band of enthusiasts who shared the sport of kings on the rough hillside or the golden strands of Connemara in the nineteen forties, fifties and sixties.

On August 6th 1951 a telegram arrived at Ross, Moyard, by special messenger. It read "Ride Rangoon Clifden today". It was signed Mc Donagh and was sent from Kilkerrin P.O. A note on the telegram in Garnet's hand reads "Rode and won the Clifden Plate". Garnet was the only girl riding amongst "the fellas" but it was not long before her skill was recognised and her services much in demand. She had the small neat figure of a jockey, she rode a tight finish and was never known to raise her stick. How was it Garnet came to be in Connemara, to be an accomplished horse woman and to be accepted as "one of us" by the flap racing fraternity?

In the eighteenth century or maybe earlier, the Irwins moved from Bonshaw on the West Coast of Scotland to the Counties of Sligo and Roscommon. Thomas

[37] S.I.M. Ch. 5 73-82

Irwin built a house in Camlin, Co. Roscommon which is where Garnet's grandfather was reared. He became Rector at Errislannan and her father, Alfred, was born there. Alfred who qualified as a Doctor, led an adventurous youth. As an officer in the R.A.M.C.[38] he travelled the world, served time in the Boer War and as a ships Doctor in the Mediterranean. This is where he met his wife Muriel whom he married in Barbados. He spent a year in the North of Russia caring for the White Russians, travelling his practice in sub zero temperatures by horse and sleigh. Some of his service time was spent at the Curragh and in Co. Meath. Garnet was born in Dublin in 1913.

When Dr. Irwin retired from the army he returned to his birth place and settled at Ross, Moyard overlooking Ballynakill harbour. Not surprisingly Alfred Irwin did not intend living the life of a retired gentleman, and for fifty years he was Doctor and trusted friend of the people of Connemara, travelling the rough unmade roads in his ancient Perry car, in all weathers, at any hour of the day or night. And indeed, throughout the troubled times when the bridge ahead of him might well have been blown up. Only those who could afford to pay were charged a fee and many a time he 'stood in' when a vet was not available. Garnet's mother enjoyed being a housewife and shared the limited social life Connemara offered. Gardening and tennis came high on her hobby agenda. She formed the Moyard Tennis Club and frequently hosted tennis parties at Ross. Alfred and Muriel Irwin and their family were soon accepted members of the local community.

It was Mrs. Lushington Tulloch[39] who gave the Irwins two Welsh Section B ponies, Rob and Roy for harness work. Garnet had other ideas for Rob and he became her constant companion until he died at the age of thirty-three. He was invariably too fat and it was difficult to keep a saddle in place but it was on Rob that Garnet taught herself to ride. Eventually she persuaded her mother to buy her a pony, one she had fallen in love with from the other side of a stone wall, a bay mare with a white blaze foaled in 1918. She cost £12.00 and it was on her Garnet learnt the finer arts of horse riding and on whom she rode her first race. Garnet, who always had an eager eye for quality, sent Swallow to the T.B. sire Buckna[40] in the hope of breeding something bigger and faster. She had no luck. In 1937 Swallow spent the entire summer with Silver Pearl and produced her only foal on the tennis court (!!) on July 1st. The filly was named Ciro. Swallow II was inspected for registration at the ripe old age of twenty with her foal at foot and was accepted for the stud book. At the age of twenty six, when petrol was short during the war, Swallow was put between the shafts and lived happily at Ross till her thirtieth year when she was put down.

Ciro was a tough little character who raced under the name of Silver Swift. She won the Connemara Derby at Oughterard in 1948 and an adult Novice show

[38] Royal Army Medical Corps. [39] S.I.M. 63-67 [40] S.I.M. 77-78

jumping competition at Galway show in 1947. In 1941 she won the class at Carna show for "Pony of Connemara type shown in saddle by boy or girl not exceeding sixteen years of age". Ciro was three years old and the only exhibit equipped with both saddle and bridle, the remainder making do with a piece of rope and riding bareback. How Garnet by-passed the sixteen age limit I'm not sure! She must have been twenty eight in 1941. None the less she and Ciro won this class three successive years taking home a handsome twenty shilling first prize to prove it. Ciro bred five foals, two to Gil,[41] two to the Arab Naseel[42] (whom she visited while competing at the Dublin Spring Show) and one to Clonkeehan Auratum. The only filly, Cilla by Gill, was the first to carry the Camlin prefix. Camlin Cilla bred Garnet three to Carna Dun before she was given to Frances Lee Norman for whom she bred three colts, the best of them I suspect, being Clonkeehan Archer. Garnet retained Cilla's only filly, Camlin Capella, who bred her three Connemara colts and finally three fillies and two more colts, the latter in Scotland by thoroughbred sires.

This very male dominated family has served a useful purpose with sires from the Camlin line travelling to Europe and the Antipodes where their qualities have been appreciated. It was a proud day for Garnet in 1953 when Cilla won the Killanin Cup and Ciro the Archbishop's Cup at Clifden show. Ciro followed this by winning her class, aged twenty-four, in 1962

Garnet kept her parents company into their ninety-fifth and ninety-fourth years respectively. When they had gone she had itchy feet and visited the U.S.A. more than once. She made good friends and even contemplated living out her days in the States. Happily this did not come to pass and she is now living contentedly at "Camlin" and is again overlooking her beloved Ballynakill harbour. She contents herself by casting a caring eye over the Park ponies as she exercises her whippet "Blue". She continues to support Clifden show where "Blue" took first place in the dog show in 1990 and 1991. She never fails to enter a hand knitted woolly in the home crafts section.

Padraig Hynes treasures a card of congratulation sent by Garnet when his mares Village Laura and Village Colleen were taking the honours at Dublin and Clifden. And more recently she sent Bobby Bolger a photo of his father Jack at Silver Pearl's head. These are clear signs of Garnet's continued interest in the ponies who were such an important part of her life.

The area of Errislannan stretches the entire length of the Peninsular from Ballinaboy to the Point. The Manor, built as a shooting lodge in 1839, nestles between a lough to its fore and a shelter belt of trees to its rear. It is the home of Don and Steph Brooks, the fourth family to live within its walls. For thirty five

[41] S.I.M. 155-156 [42] S.I.M. 195-197

years or more they have had the pleasure of calling it home, for a while as an escape from London and Harley Street but now as a permanent home.

Ponies have always been an important part of Steph's life, though she has to admit she grew up in the New Forest on a New Forest pony. I am sure one of the first things she did on arrival at Errislannan was to buy a Connemara pony.

Drimeen - An Dromin - is a "small ridge" which lies within sight of the Manor. It was here Drimeen Dun, the Errislannan foundation mare, was born in 1950. Her sire, Dun Lorenzo,[43] regularly beat the Peninsular roads alongside Josie Conroy's bicycle. Together with Errislannan Blackie (who was not born at the Manor) Lorenzo sired a daughter of great strength and hardiness who served the Errislannan stud well and faithfully over a goodly number of years. Drimeen's third foal Errislannan Daisy by Doon Paddy was the one chosen to take her mother's place to live and work at the Manor till the end of her days. Daisy won her class at Clifden in 1967 and her family of daughters and grand-children have spread the Errislannan prefix across the Irish sea to England, the Continent and Western Australia. I am not sure whether it was Steph's idea or his that Errislannan Coltsfoot covered both his dam and his half sister Daisy upon occasion, with no dire consequences I'm pleased to say.

Steph has often shuffled her cards using various sires and introducing new mare lines. For this reason there is always a wide variety of stock on view at Errislannan. The most recent departure from the norm has been the use of an Irish Draught sire on two of her mares in the hope of breeding animals with extra height and substance to carry adult trekkers.

In order to raise money to re-roof St. Flannans Church in 1967, Steph offered children on holiday a ride at 10/- a time. From this small beginning grew the Errislannan Trekking and Riding Centre which, as I write, does a busy trade in the summer months. Not content with this new venture Steph formed a Connemara Branch of the Pony Club for the benefit of the Connemara children. On a Saturday morning they arrive by the car load and all ages and abilities are catered for and encouraged.

When the Connemara Chronicle was launched Steph was the first to offer positive help. Overseas News was her idea and immediately gave the Journal an International flavour. As I researched the social history of Ireland for Shrouded in Mist, she encouraged me all the way. Together we visited the British Museum in London to turn the pages of old and precious books. Anything you do with Steph is fun and we enjoyed those excursions together.

Steph has a tremendous capacity for making everyone welcome and guess what ...... finding them a job at the same time. The beautiful garden surrounding the Manor is Don's all absorbing hobby, while the ponies remain Steph's

[43] S.I.M. 169-170

demanding pastime. They share a large family of children and grand-children who regularly find their way to Errislannan from Dublin, London, Rome or up the road from Ballyconneely. Retirement for the Brooks is a happy merry-go-round with no time for putting your feet up.

The following portraits are of two men who discovered the potential of the Connemara in the decades of 1960 and 1970. They ran almost identical operations of breeding, trading and exporting Connemara ponies, one of them in Co. Mayo and the other in Co. Dublin. They are of course John Daly and Michael O'Connell.

Lough Mask House is described as "A strong square house, bold and harsh of line, looking out Westward over lake and lawn to far off hills ... a warm secure house." *For one hundred years this has been the home of the Daly family. It was built in the nineteenth century and in its early years was used as a residence for Lord Ernes' agents. In the time of the land League War (1878 - 1881) the notorious Captain Boycott lived here and we know how he provided a new word for the English language.

'Lough Mease' is the lake between, which indicates that it lies between the Corrib and Lough Carra. It is a noble expanse of water ten miles long and four miles wide and is connected with its larger cousin, the Corrib, by an underground river. It is bespeckled by many small islands, only one of which can I find marked on a map, the Island of Inishmaine. Here lie the ruins of an Abbey, closely concerned with Eoghan Beil, the King of Connacht in the sixth century. Some of the smaller Islands have enchanting names, Inishool, Carrigeenshintilla and Carrigeendabo. The latter translated is 'the little rocky place of two cows' and has its own shred of history. It was on islands such as these that Island King,[44] Island Earl[45] and Island Duke learned their trade as sires. It was the custom to swim them to their island homes on a tow rope behind a boat. The mares were treated more gently and conveyed in a flat bottomed turf boat.

John Daly grew up on an unregistered Connemara pony and the breed was well known to him when he made his early purchases in 1960. I believe Cilcairin,[46] a Society stallion, was the first Connemara male to reach the shores of Lough Mask. His 'rambling' habits forced the Society to part with him when fourteen years old and no doubt, an island home put a stop to his rambling! By 1962 John had the four year old Island King running with a small herd. He was the first sire to carry the Island prefix. John used this prefix on his sires and that of Easter on his mares. They became very distinctive trade marks. Many stallions passed through his hands. I've found ten with the Island prefix but there were others; Ashfield Lucky Tim, Milford Hurricane, Windy Day and last but by no means

* "The Corrib Country" by Richard Hayward.

[44] O.O.M. 137-138

[45] O.O.M. 240

[46] S.I.M. 159

least, Kilgreaney Lad who spent his final years at Lough Mask. There may well be more whom I have missed, and I have not even begun to count the number of Easter females.

I first visited John in the autumn of 1981 and paid a return visit in the spring of 1992. There were few Connemara ponies grazing among the varied stock he held in 1981. His thoroughbred sire Final Problem made a lasting impression, as did an Irish Draught mare and two or three of her progeny. I do not remember seeing a Connemara stallion, though it is possible Windy Day or Island Storm were not too far away. John was pleased to share Island Duke's history with me and recalled the day he forsook his own Island and swam to an adjacent one to fight off the resident sire. Dougal always showed an intense dislike of walking through water, however shallow. Perhaps the escapade on Lough Mask left him with unhappy memories?

In 1978 Lal Berridge begged John to remove a very small, coming two years old colt who was running wild by the Screebe road and harrying the rest of the Windy herd. He was no size and backward but once he reached Mayo he prospered. As a son of Killyreagh Kim and Windy Joy he was surely well worth taking home. Registered as Windy Day he proved a brave swimmer who thought nothing of three and a half miles against five foot waves. Today he stands in Co. Clare with Harry Pendergrast. I saw some useful looking progeny by Windy Day on the 1989 inspection tour and hope breeders may seek him out. Island Storm, a son of Milford Hurricane (I wonder did he leave further sons in Connemara?) achieved Grade A as a show jumper with Alan Daly aboard, and in 1984 was the leading jumping pony in Connacht. Killgreaney Lad supplied the last sire to carry the Island prefix, Island Lad, who stands in Castlebar today.

As John developed his overseas trade he learnt the most economic means of travel and remembers sending ponies via Belfast to Montreal for as little as £52.00. One occasion remains etched in his memory. Two mares were on their way to Chicago. The St. Lawrence River froze over before they reached port. John received an additional bill 'for hay supplied from Quebec' while they waited for conditions to improve.

John was a Society Council Member and an Inspector for some years. Sadly Connemara ponies no longer grace the Islands or the park land of Lough Mask House. The era of breeding and trading the pony is past. John has a trainers licence and his son Alan is a successful jockey. I found John an interesting, genial host with horses in his very bones. Perhaps one day the Lough Mask tide will turn and encourage him to take up his love of breeding Connemara ponies once again.

Lambay Island was a perfect vantage point for marauding Norsemen and Vikings who planned to plunder the Dublin coast. An invasion of Lambay in 795 AD is recorded and without doubt it was followed by others. Lambay lies but

Milestones

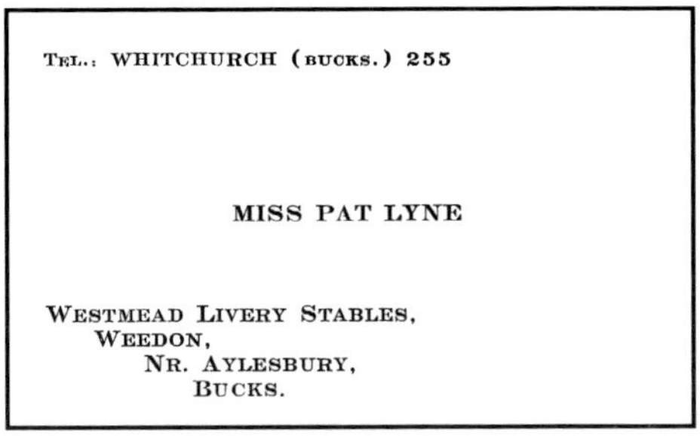

TEL.: WHITCHURCH (BUCKS.) 255

MISS PAT LYNE

WESTMEAD LIVERY STABLES,
WEEDON,
NR. AYLESBURY,
BUCKS.

Mark Geoghegan with Bridge Boy (above R)

Canrower Lass in 1968 (above)

Arctic Moon as a four year old with her first filly foal, Lune Star by Bridge Boy (below)

Milestones

Jack Bolger with Finola of Leam at Clifden (above)

Festy Mulkerrins with Atlantic Mist at Dublin (above R)

Carna Bobby, in his twenty-fourth year, at the Tulira Parade, 1970 (R)

Grey Dawn Lassie in 1970 (below L)

Breath of Spring grazing at Bunowen (below R)

Milestones

Peacockes, Maam Cross, as it was in 1971 (top L)

Mark Geoghegan (R) having a deal. Paddy looks on from behind (top R)

Jo in the side car with Mark (above)

The yearling colt by Killyreagh Kim (Truska Pimpernel) in a small stone-walled enclosure in 1973 (above R)

Ponies grazing the Truska beach (R)

The Contributors Gallery

1

2

3

4

5

1. Garnet Irwin in 1953 with Ciro and Cilla, winners of the Killanin and Archbishops Cups.
2. Steph Brooks (R.) with Errislannan Daisy, runner up to Lady Hemphill's Julie, in 1969.
3. Michael O'Connell on Lambay in 1974.
4. Lord Revelstoke with Hill Country at the R.D.S. Spring Show 1966.
5. John Daly with Easter Glory, second to Dairina for the Killanin Cup in 1957.

The Contributors Gallery

6

7

8

10

9

6. Lal Berridge fishing at Screebe.
7. Anne Hemphill with Packey Whelan and Tulira Mary Claire in front of Tulira Castle 1983.
8. Dan Llanigan O'Keefe Judging at Clifden, 1964.
9. Susan de Vesci with Abbeyleix Molly receiving the Killanin Cup in 1967.
10. Stanislaus Lynch on Good Heaven by Carna Dun, 3rd in the unmade Heavy-weight Polo Pony Class, Dublin Horse Show 1961, leading his mare, First Flight, Reserve Champion Brood Mare of the Show.

The Contributors Gallery

11

12

13

14

15

16

11. Jane Dalrymple (L.) with Bridget Snow at the launch party for Out of the Mist, Clifden 1991.
12. Frances Lee Norman with Clonkeehan Tiger Lily receives her challenge Trophy from Jan Harald Koelichen in 1965.
13. Lib Petch receiving the Lady Carew cup with Coosheen Alanna in 1963.
14. Jimmy Jones presents 'Best in Show' Trophy to Outi Talvi with Brunskaer Marguerita at the Finnish September Show, 1991.
15. Lady Maria Levinge in 1990.
16. Claire Studdert with her yearling filly by The Fugitive, a winner at Clifden in 1975.

The People's Gallery

17

18

19

20

21

23

22

17. Jim Walsh, 1992.
18. Johnny Lee at the Inspections, 1990.
19. Jimmy Canavan and his son John, 1992.
20. Bobby Bolger at Derrygimla, 1992.
21. Mark Geoghegan in 1977.
22. Henry Whyte 1992.
23. Jack Bolger with the prize-winning mare, Heather Mixture in 1966.

The People's Gallery

24

25

26

27

28

29

30

24. Padraig and Mary Hynes at the door of their new home, 1992.
25. Joe Gorham, 1992.
26. An unidentified face at Clifden Show 1984 - Josie Ridge.
27. Henry O'Toole, 1992.
28. Josie Conroy, 1992.
29. Paddy King at Lehid 1993. (Paddy Conneely was out fishing the day this photo was taken).
30. The Festy Kings with Power Station in 1992.

The People's Gallery

31

32

33

34

35

36

31. Thomas Burke at Clifden Show 1963.
32. Stephen Keaney with Winnie Love at Cashel, 1992.
33. Michael King of Derryeighter, 1992.
34. Mikey King of Cushatrough with Mervyn Kingsmill, receiving his Trophy from Michael O'Kennedy, 1986.
35. Graham Tulloch with Loobeen Larry, 1989.
36. Petie Molloy and his son with the Dooneen ponies, Clifden 1986.

The People's Gallery

37

38

39

40

41

42

37. Colman Griffin at Maam Cross, 1978.
38. Paddy Folan with May Harp, 1992.
39. Tommy McDonagh with Dairina, winner of the Killanin Cup, Clifden 1957.
40. Willie Diamond (R.) presents a Trophy to Joe Murphy in 1983.
41. James de Courcy with Cuchulain, winner of the Stallion Class 1986.
42. Noel Mannion, 1991.

The People's Gallery

43

44

45

46

47

48

49

43. Mark Cloherty and Winning Ways, 1992.
44. John Luskin with his prize-winning Stallion Milford Hurricane in 1977.
45. Murty McGrath with Anne Hemphill in 1991.
46. Michael Clancy.
47. Joe McNamara at Oughterard Show with Ganty Gina, 1989.
48. John Conneely beside the cart drawn by Ballydonnellan Grey.
49. John Enda Feeney.

The Steward's Gallery

50

51

50. Bartley O'Sullivan.
51. Michael O'Malley.
52. John Killeen.
53. Phil McDermott.
54. Sean Keane.
55. Father White.
56. Donal Kenny.
57. Lord Killanin.
58. Father Cunningham.

53

54

55

52

56

57

58

three miles off the Dublin coast and was formed by volcanic eruptions at the close of the Silurian era. The famous green porphyry of Lambay is the home of Lord Rupert Revelstoke, manx sheerwaters, peregrine falcons, a herd of deer and for fifteen years, 1962 - 1977 the world renowned Lambay Connemara ponies.

Lambay was bought in a state of dereliction at the turn of the century by the Hon. Cecil Baring. The ruins of a Norman Castle - more invaders? - and stone out-buildings were restored and transformed into an English style manor house with the help of Sir Edwin Lutyens, a personal friend of Cecil Barings. The original gardens were designed and planted by Gertrude Jekyll.

It was here Michael O'Connell came in 1952 as farm manager for the present Lord Rupert Revelstoke. Michael, a keen horseman from his earliest years, (spent in West Limerick), studied Agriculture at University College, Cork in the 1940s. It became his habit to travel to Connemara every autumn to buy store cattle to graze the Island and thus he got to know the people and the region which, he discovered, enjoyed similar climatic and grazing conditions to that of Lambay. This encouraged him to introduce Connemara ponies to the Island. He bought weanling foals at the same time as the cattle and allowed them to graze and grow in freedom, until three years old when they were broken and sold. Michael had six children, so no shortage of jockeys. Lambay extends to one thousand acres, only three hundred of which are good farm land. The remainder is unfenced rough hill grazing which rises to 315 ft. above sea level. It provided an ideal area on which to rear Connemaras.

A Carna Dun filly, Lambay Star, registered in Volume IX of the stud book, was the first pony to carry the popular Lambay prefix. One hundred and fifty ponies, or more, were subsequently registered as Lambay ponies, though it is thought double this number passed through the Island herd on their way to new owners.

An instance of the misuse of a prefix in those days was highlighted by Michael in 1967. He wrote to the English Connemara Quarterly to correct a reporters error in the Riding magazine which stated that Lambay Rebel,[47] Lambay Inver[48] and Lambay Kate were bred by Lord Revelstoke. None of them were bred on the Island but were purchased as foals. At registration time, when two years old, they were entered in the stud book as "Lambay" ponies the property of Lord Revelstoke but with the rightful breeders name recorded. Michaels public recognition of what was a fairly general misuse of prefixes in this period, was welcome.

In 1964 Lambay Rebel won the stallion class at Clifden and two yearling colts by Camlin Cicada, shown the same day, gained 1st and 4th place in their class. Lambay Rebel was the Islands first resident sire, home reared if not home

[47] O.O.M. 188-189

[48] O.O.M. 224

bred. The two colts returned to Clifden in 1965 as Lambay Fireball and Lambay Inver. They arrived too late for their class and were not judged. What a disappointment after the long journey from the Island. Lambay ponies were rarely, if ever, catalogued again.

As the price of foals increased it was decided to switch to breeding, and mares by Gael Linn, Tiger Gill[49] and Paddy's Heaven were bought. The herd increased over a period of time and several good sires were used. Murrisk,[50] Ben Lettery,[51] Fort Lambay and Moy Village Master were but a few. Michael's stock were always on the move and there was no suspicion of line breeding or inbreeding. His favourite sire, Lambay Bobby, a son of Carna Bobby was bought from his breeder Joe McNamara as a foal. While, on principle, every pony on the Island was for sale Bobby was treated as an exception and died on Lambay at a comparatively young age from an unknown cause. Michael built up a brisk continental trade and Lambay ponies were prominent in the early stud books of several overseas societies. Unlike many of his counterparts he was a good correspondent and the contacts he made flourished as a result of his letter writing.

Michael's knowledge of the breed and his enthusiasm for it was welcomed by the Connemara Society and he was soon playing a prominent role within its life. In 1966 he was asked to Judge at Clifden in company with Tim Cotter, T. McD.Kelly, R. Kelly and Dan Llanigan O'Keefe. It was Tim Cotter's last year after nearly forty years on the Judges panel. Michael judged at Clifden on a further six occasions between 1967 - 76. By 1972 the Society's Inspection Committee required consideration when the original long standing team retired. Michael was one of six new names added to their list. He became a valuable and a very active member of the Society and his death in 1977, at the early age of fifty-four, was a grievous loss to them, as well as to his family and friends.

It was sad indeed that the Lambay herd was dispersed on Michael's death. Many of the ponies, the cream I suspect, were sold privately. The remainder were taken to the Ashbourne Mart on April 14th 1978. My old friend George Mullan, by now a very sick man, but, as ever, ready for a deal, rang and suggested I fly over for the auction. Inevitably it was rather a sad occasion to see ponies from eleven months to fifteen years being sold under the hammer. I bought a six year old grey mare by Ardilaun, Grey Crayon, for 240gns. She was carrying a foal to Moy Village Master, one of two stallions sold the same day.

I have not had the pleasure of visiting Lambay or indeed of meeting Lord Revelstoke. I have had to content myself with an unequalled view of the Island from the Carmel at Malahide which is situated on a hill opposite, overlooking the bay below and Lambay beyond. I am grateful to Eithne O'Connell, Michael's daughter, for her kind help.

[49] S.I.M. 158-159 [50] O.O.M. 39-40 [51] S.I.M. 174 + Index O.O.M.

Lord Revelstoke has encouraged my writing with a postcard in 1984 and a poem, written by himself in 1992.

DID THIS FAMOUS BREED COME HERE BEFORE THE DISTANT DAWN
OF CHRISTIANITY? - If so, the strain is just as old
As the Celtic race itself ... or was this noble pony born
In fertile Andalusia - escaping here, we're told

From ships of the Armada wrecked along these western shores...
Or were they brought by merchantmen (whose Galway City wares
Were traded then by Spaniards from Madrid to the Azores?)
Or were they gaily interbred with native Irish mares?

Whichever tale is true, this simple fact remains today -
This hardy breed of horse withstood this land of grass unsown,
Of mountain, lake and bog and gorse - rough herbage (lack of clay)
Unshielded from Atlantic storms except by rock and stone.

On Lambay's eastern sea-board, wind-swept cliffs are much the same
And "hardy, strong and free" these ponies thrive with healthy grace
Displaying charm of manner which equestrians acclaim...
And happily for us they show, how much they love the place.

R.

I am sorry I did not know Stanislaus Lynch. I have a feeling we would have enjoyed each others company. He was a man of diverse talents whose Literary achievements included the award of a Diploma for Epic Literature. He received this award in the Arts Section of the 14th Olympiad London 1948 for his "Echoes of the Hunting Horn", one of his six published works. His biographical notes tell me he had two books awaiting publication one of which was "The Incomparable Connemara". I wonder what became of this?

Stanislaus Lynch was born in Co Cavan and educated at St. Vincents College, Castlenock, Co Dublin. During his lifetime he hunted with more than one hundred packs of hounds, across Ireland and in France and America. While living in Ballyjamesduff he kept a private pack of hounds which he hunted himself. He regularly exhibited at the R.D.S. Show and in 1961 his mare "First Flight" was Reserve Champion mare of the show (Coote Cup). On the same day he rode an overgrown Connemara gelding "Good Heaven" by Carna Dun into third place in the unmade Heavyweight Polo Pony Class. Stanislaus was the complete horseman.

As a Journalist he was a regular contributor to The Field, Horse & Hound, Le Saint Hubert (France) L'Anee Hippique (Switzerland) Hoofs & Horns (Australia) and every aspect of the Irish Press. He was renowned for his running commentary for RTE on the World Show Jumping championships at the R.D.S. He contributed poetry readings for the same medium and was called upon to lecture from time to time.

And yet it was the Connemara pony who played such an important part in the life of this interesting man. In 1953 he was invited to undertake a tour of Connemara studs in the U.S.A. and travelled from the Canadian border in the East to Georgia, Alabama and as far West as Texas. In 1959 he sent fifty four Connemara ponies to Mr. R. H. Wright of Rose Hill Farm, Georgia, I notice his printed notepaper has "Irelands Largest Exporter of Connemara Ponies" at its head.

In 1960, sales to the President of India were followed by others to the continent. In Volume XI of the C.P.B.S. stud book, Stanislaus Lynch had thirteen ponies registered in his name and in the same Volume John Daly had nineteen entered and Mr. Wright of the U.S.A. ten. Lord Revelstoke had only one, but by Volume X he had twelve and Mr. Lynch eight.

Stanislaus did not adopt a prefix but was a genuine breeder with his own stallion and some resident mares. It is obvious stock were 'bought in' to fulfill his overseas orders, many of whom were exported before registration age.

I am interested in his apparent penchant for blue eyed cream ponies. It was he who bought the blue eyed cream colt Marconi after he won the yearling colt class at Clifden. Marconi[52] was later exported as a stallion to Mr. Wright of Georgia, but was used by Stanislaus before he left. He retained two of his sons as entires, Lynch Law and Spanish Arch both of whom were grey. He also bred from a blue eyed cream mare, Snowflake, who reproduced her own coat colour on at least one occasion before travelling to England where she was greatly valued. I wonder if Stanislaus had his own theory about breeding from these ponies?

Margaret Lynch, Stanislaus' widow, lives in Co. Meath and keeps one Connemara mare and thus retains an interest in the breed. She, it is, I have to thank for the material in this Cameo.

It is apparent that Stanislaus Lynch, John Daly and Michael O'Connell were responsible for creating an interest in the Connemara pony around the world. It was they who purchased the stock from the Connemara heartland and developed the overseas trade. It would hardly have been possible for the pony-men of Connemara to do this for themselves in those early times. Lynch, Daly and O'Connell were men with an experience and knowledge of horses way beyond the Connemara pony. I'm sure they rendered the breed and the people a service, though critics may feel they bled Connemara of some of their best stock. At the same time it has to be said a few of those exported did the breed no favours. This was

[52] O.O.M. 104-105

inevitable when large numbers were leaving the country, and hopefully the gains were greater than the losses.

Tulira Castle and many of its inhabitants are cloaked in colourful Irish history. The Norman Tower was built by the De Burgos in the thirteenth century. Between the 15th and 18th century it was extended by court-yard wings. The main castellated house was built in 1882 by Edward Martyn and was designed by Ashlin, the best Gothic architect of the time. The castle passed into the hands of the Martyn family sometime before the end of the sixteenth century. This family was established in Ireland about 1170 by one, Oliver Martyn, a Captain in Strongbows army. Later he fought in the Third Crusade distinguishing himself in Palestine and becoming the "Right Hand Man" of King Richard the First of England, better known as Richard the Lionheart. After being shipwrecked in the Adriatic on their return journey, they were imprisoned together in Austria. Oliver Martyn died in captivity but the King, who ransomed himself, granted the famous coat of arms to the Martyn family on his return home. The coat of arms, one of the earliest, is a cross surrounded by the Sun and Crescent of the Moon with the star a the crest signifying "Glory" which is reached by way of three "Lights" i.e. Christianity, the Sun and the Moon. His descendants in Galway became known as one of the fourteen ancient families, of "the Tribes of Galway".

In the twentieth century Edward, a celebrated member of the Martyn* family, lived the life of a typical Galway landlord. He rode to hounds, was a good shot, sat on the Galway Grand Jury and was a practising Christian. Besides this he was a literary man of quality who founded the Abbey Theatre and was a financial supporter of the Irish Literary Revival. He wrote plays, had a natural ear for music and entertained lavishly. It was he who created the splendours of the Castle, the leather covered walls, the Irish marble fireplaces, the stained glass windows and the hand painted ceilings. And yet it was his choice to live, write and die in the Norman Tower, which was, I believe, austerely furnished. He died a bachelor and in 1923 was succeeded by his first cousin Mary, Baroness Hemphill (nee Martyn). The present Baron, Peter Patrick Hemphill, moved to Tulira with his wife Anne and their small daughter in 1952. They are the subject of my next cameo.

Lord Peter Patrick is a countryman who enjoys sailing and all field sports. In 1987 he held the prestigious post of senior steward of the Turf Club. He shares his wife Anne's love of fox hunting and had three terms as Master of The Galway Blazers, 1957-1960, 1967-1969 and 1972-1985. Together they founded the Tulira Connemara Pony Stud in 1963. For twenty years the ponies grazed the Park and farm land of the Castle and a herd of a distinctive type was soon established.

* Martyn, a Norman name, was spelt with a Y.The anglicised version, Martin, substituted I for Y. Edward Martyn chose to revert to the use of Y. The Martins of Ross and Ballinahinch were also descendants of Oliver Martyn.

Anne Hemphill (née Ruttledge) grew up in Co. Mayo on the shores of Lough Carra near Ballinrobe. She was an avid foxhunter from her earliest years. Her first pony, Peggy, was reared at Screebe, cost £15.00 and was bought on the advice of Paddy Daly, father of John. Peggy was an all purpose pony who took foxhunting, show jumping, pulling the trap and breeding to a thoroughbred sire in her retirement years, in her stride. In the war years Anne travelled thirty miles to the nearest meet of the Blazers. She hacked five miles to the railway station where the pony was loaded with a change at Claremorris station being necessary. She recalls her frustration at having to leave before hounds were blown home in order to catch the last train. Some years later Anne was Field Master of the Blazers for fifteen seasons. She founded the Galway Branch of the Pony Club and was District Commissioner for twenty five years. It is not surprising the Tulira ponies are renowned for their strength and boldness. They are the type of pony you would expect the Hemphills to breed, ponies capable of doing some real work and remaining sound.

The stud was founded on two mares bought from the film director John Houston, who lived nearby at Craughwell, Glen Nelly and Star of Fahy. A son of Glen Nelly, Tulira Mairtin was used as foundation sire and the mares Noreen Grey, Glencaugh Star and Julie were bought to run with him. He proved a very dominant and fruitful sire. Many of his strong-boned sons stand in England and elsewhere on the continent. He is well represented in his homeland with three sons within a stones throw of his old home, Tulira Castle. They are Smokey Playboy at Raford, Tulira Guinness at Craughwell and the 1991 Champion stallion Tulira Finn MacCool at Claregalway.

I consider myself lucky to have visited Tulira Castle on more than one occasion. An overnight stay in 1980 remains a clear memory; the large open log fires in the hallway and every downstairs room through which we passed were surely one mans job to stoke and keep alive; the magnificent broad marble staircase which my old sheep dog Tansy found so difficult to climb; and my own struggle the following morning with the massive castle door and key (which turned back to front) when Tansy asked for a run at an unsocial hour. A chore, I may say, which was more than repaid by the beauty of the Park and its Castle in a rising sun.

It was a sad day when the Hemphills decided to leave Tulira and disperse the pony stud in 1983. It is a strange irony that ancestral homes such as Tulira Castle can no longer be maintained by those to whom they rightly belong.

Anne remains an energetic and resolute lady and in 1989 re-founded the Tulira Connemara stud at Raford where the Hemphills now live. She was determined to find mares of her own breeding for this new beginning and she was lucky to do so. The Tulira prefix is well and truly back on the Connemara map.F

Baile Na hInse, the settlement of the island and Ballynahinch to most of us, is steeped in history coloured by the people who lived there. A stronghold tower was built on the lake-island by the O'Flahertys in the 13th century and was fiercely fought over by various branches of the clan until the times of Cromwell. The O'Flaherties established a Carmelite Monastery, and it is possible that the first house built by the Marins, three centuries later, was on the same site. Described as an Inn it was replaced by a modest house of the style of the 1660's but given the title of Castle. The Martins presided over the Barony of Ballinahinch, which covered some 160,000 acres, from 1714-1849. "Nimble Dick" was followed by "Humanity Dick"[53] and this well chronicled family, a member of the notorious Tribes of Galway, were feudal landlords renowned for smuggling, duelling, kindness to animals, extravagance (stables built lined with marble) and quite unusual hospitality. Maria Edgworth's visit in 1833 is one I cherish. Determined, as she was, to meet the "King" of Connemara she ventured into his wild and unruly kingdom in her sixty-sixth year. She was travelling the newly built roadway from Galway to Clifden and as she approached Ballinahinch the road subsided into the bog under the weight of the coach. Rescue came from nowhere as thirty assorted men and boys emerged from the trees and lifted the coach and its passengers bodily onto dry land and saw her safely to the Castle gates. In 1883 the Martin fortunes were declining and Maria found the once elegant house and furnishings sadly tarnished. By 1847 the Martins were bankrupt and the estate bought cheaply by Law Life who refurbished the house inside and out adding battlements, to justify the title Castle maybe? In 1872 it was resold to the Berridges who later added a third storey and a terrace to the house. This is where Lal Berridge grew up.

Born in January 1910 and christened Eulalia Beatrice she had her own pony from an early age and rode the woods and lanes of the Ballinahinch estate. The mountain top, Ben Lettery, kept a stern eye on all her movements. In 1914 the Congested Districts Board bought the bulk of the estate and divided the land among the tenants. The Berridges retained the Castle and townlands of Ballinahinch and Killeen until 1925 when they sold their remaining interest in Ballinahinch and moved to Screebe Lodge. Lal had many happy years there and managed the fishing and shooting interests until the early fifties. She told us tales of fishing sharks by night and enticing seals to follow the boat by day to the music of a gramophone. These were halcyon post war years and I wonder what persuaded her to leave Connemara and settle in Co. Wicklow?

In 1959 Lal bought Mervyn, an attractive house in Enniskerry, and it was here she established her Connemara stud. I guess she could not live outside Connemara unless she had some part of it with her. She maintained close links with Connemara and the Pony Society. She was a respected judge and inspector for the

[53] S.I.M. 14-15

Society for many years and her Mervyn prefix soon became one to be reckoned with and sought after.

I remember Lal came to England in 1968 to Judge at Pearl Lawson Johnsons Connemara Rally at Melchbourne. Lal was judge and I exhibitor and we did not meet on this occasion. At the International Conference in Galway in 1970 she was called upon to report on the life of the Connemara Pony Breeders Society from its formation to the present day 1923-1970. It was a simple, factual report but ended on a firm note "In conclusion may I remind you the Connemara Pony Breeders Society is not only the parent pony Society, it is also the cradle of the Connemara pony of today. You all like our pony, and it is OUR Connemara pony, you like it enough to breed it in your own countries. Accept it as we have produced it. We know it is the very best pony in the world."

On our round trip in 1977 Mervyn was our last port of call. Lal invited us to lunch. In the event we spent the entire day with her helping to prepare an evening meal after bringing the mares and foals in ready for weaning the following day. Jo spent most of her day meditating on a Renoir pastel portrait hanging on the study wall, "Better than lying in a bank vault" was Lal's response to Jo's incredulous surprise and joy at recognising such a treasure. We found Lal a naturally reserved person with a lively sense of humour. The day passed all too quickly and I am sorry it was never repeated.

Lal founded her Mervyn stud on two mares bred in Connemara, Windy Rebel and Winter Roche.[54] Lal and Colman Griffin were good friends and I believe Windy Rebel by Lavalley Rebel was the best of Windy's many daughters. As we have already seen, more than one breeder was tempted to use Naseel in his heyday and Lal was another. Naseel repaid her with two good fillies, the first of whom, Mervyn Mistral, provided the breed with Mervyn Pookhaun,[55] Tom and Liz Ormsby's attractive sire by The Admiral. I remember greeting Lal with excitement and congratulations when he won the stallion class at Clifden. Her tart reply was "Your society in England would not have him in their stud book". At the time she was very unhappy at the new rules introduced by the E.C.P.S. I could find no answer but some years later when our rules were re-written I tried to lease Pookhaun from the Ormsbys (without success) and that is a measure of my regard for this pony. Camlin Cicada and Windy Rebel provided two good sires in Mervyn Storm and Mervyn Invicta and a daughter Mervyn Mayfly. Lal owned the stallion Abbeyleix Cypress for a while and later used a colt of her own breeding, Mervyn Bowsprit, on Windy Rebel. It was a daughter of Cypress and Windy Rebel, Mervyn Blue Charm who provided the society with Mervyn Kingsmill. I am sorry Lal did not see the full fruits of his years in Connemara before she died in 1988.

[54] S.I.M. 220-221

[55] O.O.M. 279

She would have been justifiably proud of the contribution he has made to the breed.

Lal's ponies gained a considerable reputation at the Royal Dublin show both in the Connemara ring and the jumping arena. She encouraged young people to ride and produce her ponies for her, which they did with success.

In 1992 I paid a visit to the graveyard behind the wayside Chapel at Ballinafad where Lal is buried alongside her parents and one of her brothers. Ben Lettery continues to keep his watchful eye on her. May she rest in peace in the knowledge that the Connemara pony will always recognise her place in its progress.

Dan Llanigan O'Keefe (née Malcomson) was born in 1913, the same year as Garnet Irwin. Unlike the self-taught Garnet, Dan grew up in a family steeped in horses and was out hunting at an early age aboard a donkey. Throughout her life she hunted with the Meath, Kildare or Ward Union Foxhounds and had a close family association with the Tara Harriers. This was the breath of life to Dan, as was point to point racing, a tough sport for a girl in the days of her youth.

Eileen Amenia - no one is sure where Dan came from - married Randal Stephen Llanigan O'Keefe in 1935. They had two children John and Susan. Dan enjoyed a great love of the West of Ireland which she visited regularly, but she did not become actively involved with the Connemara pony until her children reached riding age. They were mounted on Connemara ponies and were soon enjoying every kind of sport with them, from hunting to the show and jumping rings of Dublin. Glencara Pet, a Dun Lorenzo mare, is one remembered as 'special' by Susie.

Dan was born with a natural gift, an eye for a horse, and her greatest talent was as a judge. When still a newcomer to the breed I recognised her profound knowledge of it. We invited her to stay at Redgates after she judged at the National Pony Show in 1978. I still have the notes I made of our conversation that evening. She spoke of the beauty of a herd of Connemara ponies grazing peacefully in their own pattern, merging into the background of their native heath. And the thrill of seeing first one head and then another rise, as if startled by something on the wind, before gathering swiftly and galloping into the distance beyond the skyline, to find another sheltered haven. Dan would deeply regret how rare a sight this is in Connemara today.

Dan did not consider herself a breeder, though she was moved to rescue the mare Rambling Colleen in 1965. She noticed her rearing a foal year by year on the same roadside with little care. When she felt she might not survive another winter in Connemara she found her owner and bought her.

During the 1960s and 1970s Dan was a regular Judge at the Clifden Show. In the 1980s her daughter Susie appears to have taken her place. Bent Nielsen came to rely on her judgement and she travelled to Denmark every year at inspection time in the 1970s and Judged at the Roskilde Festival in 1977. In 1970 she was invited to Judge in the U.S.A. She suffered a severe culture shock and hardly recognised some of the ponies brought before her as Connemaras. The Americans were very hurt by her outspoken comments. I'm sorry they did not understand each other better.

In my diary an entry for Oct. 10th 1981 reads: "I spent the night at Tara with Dan and Llani. We had trout for supper (caught by Llani that morning) in front of a log fire and washed it down with red wine. The evening passed quickly, raking through old photographs and memories. Dan and I have so much in common. By 1.30 a.m. it was too late to return to Bray. I went to bed in the spare room in a pair of Llani's pyjamas, with Tansy at my bedside."

Dan died in November 1990 and with her death an era passed. She was only a little lady but a great one, with a fount of knowledge and a true understanding of the Connemara pony. She liked to remember him as he was in the post war years, when he emerged as an all purpose riding pony but with his native characteristics intact.

"On the outskirts of the town of Abbeyleix[56] is the beautiful demesne of Viscount de Vesci with a Wyatt Mansion of 1774 in which can be seen, near the traditional site of the ancient Abbey, the tomb of Malachi O'More, Prince of Laois in the 13th century."

In 1962 Susan de Vesci and her husband visited Clifden show and together they fell in love with the Connemara pony. This was newsworthy and included in the Connacht Tribune show report.

In the parkland of the Abbeyleix estate Susan ran her herd of Connemara ponies between 1962 and 1983. Our visit in 1977 is yet another happy memory. I recall the ancient oak tree with a bed of miniature cyclamen at its feet; the well matched grey mares and their foals, who mingled contentedly among the trees; the beauties of the house and its contents with Susan in the kitchen preparing the lunch herself. Abbeyleix is one of Irelands many historic and beautiful homes, whose maintenance is a constant struggle for their owners. Susan endured this struggle cheerfully and the ponies were her joy and therapy in a busy life.

In Volumes IX - XII of the stud book nineteen ponies were registered with the Abbeyleix prefix. Only one of them was bred at Abbeyleix. In common with many breeders of the period the use of a prefix was not restricted to those she bred herself. As time went by Susan became a very discerning breeder and assembled a small herd of brood mares, all of a type. The Abbeyleix prefix soon became a valid and much respected one. Abbeyleix Molly (reputed to be her favourite), Rebel

[56] Description from AA Gazetteer 1962

Lass, Finola of Leam and Callowfeenish Dolly II all contributed to the success of the stud. Susan did not stand her own stallion but bred more than one good one during the years. She was not afraid to use the unfashionable sire or to travel long distances in order to do so. Her ponies achieved considerable success in the show rings of Dublin and Clifden, crowned with the Supreme Championship at Clifden with Abbeyleix Delphinium in 1983. Susan and her ponies were always welcome in Connemara. She was a gracious, gentle lady and the pony men respected her. I believe her eye for the pony grew with the large number who passed through her hands. She was invited to Judge at both Clifden and the Breed Show in England.

When Viscount de Vesci died Susan's herd was dispersed and she moved to England in order to leave Abbeyleix to her son. This was a great sadness for her but she never complained. I remember her joining John and Phyllis Meade and myself for breakfast in the Clifden Bay Hotel on show morning 1985. She was already seriously ill by then but you would not have known. She was full of excitement for the day ahead.

Lib Petch was Susan de Vesci's good friend. They began breeding Connemaras at roughly the same time and Lib continues to do so, albeit on a smaller scale. A report in the Irish Farmers Journal, after Clifden in 1965, makes interesting reading twenty-seven years later; "Miss Lee Norman's success was further evidence of how seriously Connemara breeding is being tackled in areas outside Connemara. Less spectacular but equally impressive were the two rosettes which Mrs. E. Petch took home to Kilbrittain, Co Cork, wife of a very progressive young farmer, John Petch, she has built up a herd of ten pure bred mares and this was her first time to show Cork breds. The rosettes were for third places in classes for yearling colts and yearling fillies. They hadn't been sold and I'm not sure they even were for sale but Mrs. Petch told me she had recently sold two four year olds for export to Sweden at £140.00 ex farm".

Lib has always run her own stallion. The first was Ceilog, a son of Auratum, who won his class at Clifden in 1962. For a while she leased Errislannan Coltsfoot before buying Coosheen Finn, a son of Carna Bobby and Finola of Leam. One of her early purchases was Coosheen Alanna to whom I owe an apology. It was she and not Arctic Moon who won the Clifden two year old class in 1963*. These two good fillies were, strange to say, next door neighbours in the stud book as well as in the show ring. Alanna was subsequently sold to Lord Hemphill and became a brood mare for the Tulira stud. Two of the Coosheen stud's foundation mares were daughters of Cregg Lassie and Silva,[57] Coosheen Petronella and Sylvan Vale. They have bred Lib some good ponies and she has been repaid for remaining loyal to her female families. Many a time the long journey from Cork to Clifden has

* See O.O.M. P.91

[57] S.I.M. 225-226

been worthwhile, never more so than in 1979 and 1980 when Coosheen Nutmeg won the Championship two consecutive years, a unique achievement.

I am sure it gave Lib much pleasure to own one of the "great" mares of Connemara, Village Belle,[58] for the last years of her life. She was blest with a daughter and grand-daughters from this illustrious line. Lib has been generous in her travelling time and a good ambassador for the breed when judging overseas.

Frances Lee Norman could be an intimidating lady but she had a heart of gold. In many ways she was larger than life and her indomitable spirit carried her through when arthritis crippled her later in life. Frances was born in Belfast but grew up in England in Devonshire. She enjoyed all the country sports as a girl and in 1939 joined the F.A.N.Y.S.* in which she served with distinction throughout the war. After the war she bought a property at Slane in Co. Meath and there she founded the Clonkeehan stud. Frances was a true all rounder who enjoyed the countryside, farming and everything to do with the horse or pony. The Clonkeehan stud was soon firmly established and ponies carrying her prefix were often in the Championship spot at Dublin and Clifden. She was a keen business woman and exported good ponies throughout the world. She had the facilities and bred regularly from her foundation mares Nipper of Calla and Rose of Barna. I notice she occasionally used Clonkeehan as a suffix. All those carrying the prefix were genuinely bred at Clonkeehan. She experimented with cross breeding using the renowned Arab pony sire Naseel on two of her Connemara mares. She also used her Connemara sire Clonkeehan Nimbus across thoroughbred mares with considerable success. Her able assistant, Susan Trail, was an important cog in the busy life of the Clonkeehan stud.

On a round trip home from Connemara in 1977 we stayed two nights at Clonkeehan where we were made most welcome. The coat colour study Frances and Susan undertook on the progeny of sires in the first twelve volumes of the stud book was of immense value to me as I worked on Shrouded in Mist.

Clonkeehan Auratum will ensure Frances and her contribution to the breed will never be forgotten. For all the controversy his place on the C.P.B.S. stallion list may have caused, the reality is, that the stud book would have been the poorer without him.

Frances was invited to Judge at Clifden in 1959, no better a one to be the first lady to stand in the hallowed centre of the Clifden ring. She became a member of the Society inspection team in 1972. The Clonkeehan Challenge Trophy is awarded every year to the best pony, bred in and owned by, a resident in Connemara. She knew how it felt to come in from 'outside' and deny the Connemara men the coveted championship trophy, and the presentation of the Clonkeehan Cup was a typically thoughtful gesture.

[58] O.O.M. 76-77

* First Air Nursing Yeomanry

Jimmy Jones owns a shoe shop in Carlow but I suspect he spends more time in his pony pursuits than in his shop. His Ashfield prefix does not appear in the stud book until 1970 when Wise Sparrow provided his first progeny. Good mares such as Atlantic Surf, Roundmount, Nimble Fingers, Garrafin and Willow have joined the descendants of Wise Sparrow at different times. Many of the Society sires have spent the last years of their life with Jimmy, MacDara, Atlantic Storm and more recently Killyreagh Kim and Ormond Oliver.

It became Jimmy's hobby to travel Connemara and to learn the strengths and weaknesses of the various blood lines and there are few pedigrees he does not know off by heart. He works very hard on behalf of the Society at inspection time and as a Council member.

Clair Studdert and her brother Bob are, she told me, 'blow-ins' for they moved from Co. Clare to Co. Meath in search of better farm land. Littlerath, Trim is where they settled and here they enjoy their separate horse and pony breeding interests.

Clair bought her first Connemara pony, Easter Song, as a two year old in 1962. She was foundation mare for the stud and is remembered in perpetuity with Song appearing in the name of every pony bred at Littlerath. When I called in 1992 she was represented by a seventeen year old daughter Song of Regine, a grand type of flea bitten grey mare by Clair's resident sire The Fugitive.

In 1969 Charlie Lydon, a Veterinary surgeon from Westport, bought a colt foal by Kimble from his breeder and named him The Fugitive. He impressed Clair at an early age and as a yearling she did a deal with Charlie and took him home to Co. Meath. In the year of the Conference, 1970, he was two years old and travelled to Westport the day before the show. Clair drove up on show day, having arranged to meet him at Clifden. She broke down on the way, thumbed a lift on the back of a motor bike to the nearest garage (nearly died of the cold!) and just made it to the show on time. What an impression The Fugitive made that day! He stood second to his sire Kimble in the senior stallion class and then won the two year old class. How brave of Charlie and Clair to 'take on' the older stallions, at the tender age of two. The Fugitive has gone on to sire many good stock and as he approaches his twenty-fifth year perhaps it is a pity there are not more of them.

Clair has never lacked initiative and towards the end of 1970 she imported a Danish bred sire Oxenholm O'Connell. He has since been replaced by Billy, a four year old son of Catamaran, who shows plenty of pony character and should go well on her Fugitive daughters.

Maria Levinge (née Carraciola) fell in love with the Connemara pony on her first visit to Co. Galway as a child in 1957, when she and her mother enjoyed a holiday at Screebe Lodge, (rented from the Berridges). Maria was born on The

Island in Waterford and her mother - who was a Fitzgerald - was a keen foxhunter who rode to hounds side saddle. Her father was an Italian aristocrat. Maria never ceased to be grateful circumstances decreed she was brought up in Ireland rather than Italy.

Maria's foundation mares were Solitude by Clonkeehan Auratum and Dun Sparrow by Carna Dun. When the stud was established in 1961 she was living in Grange Road, Rathfarnham, Co Dublin and Grange became her prefix. Later she moved to Co. Wexford where in 1978 she met and married Sir Richard Levinge. The Levinge family came to Ireland in the late 17th century when the first Sir Richard was appointed Lord Chief Justice to Ireland. Clohamon is not the original family home and Richard has no 'in-born' interest in horses or ponies but now accepts them happily as an important part of life at Clohamon. Since her marriage to Richard, Maria's breeding programme has grown and prospered in the comfort of the Clohamon paddocks at Bunclody. While remaining faithful to her foundation mares to the end of their lives, she has introduced to her stud some of the oldest and best mare lines from the heart of Connemara. Merry Dolly, from the Carna Dolly[59] family, Kilkerrin Surf by Thunderbolt,[60] Home Lover by Clonjoy and most recently Winnie Love the great grand-daughter of Winnie.[61] There was a field full of promising youngstock when I called in 1992 and they will keep coming while Callowfeenish Mairtin continues to rule the roost. He does so with great intelligence and has shown a high fertility rate. Mairtin, a son of Abbeyleix Owen,[62] has every opportunity to prove himself a high class stallion with some of the best mares visiting him from far and wide.

Maria's busy life at Clohamon has made it necessary for her to reduce her commitment to the C.P.B.S., but she continues to find time to visit overseas Societies, when invited to share her knowledge with them.

Jane Dalrymple has a small and select stud at her Boden Park home in Co Meath. This is where the mare Kingstown[63] spent her last breeding years and her progeny were the first to carry the Boden Park prefix. She was joined by a mare very different in type, Clonkeehan Whitethorn, who is out of a Naseel mare Clonkeehan Mayfly. When I visited Boden Park in 1990 Whitethorn did not look her twenty six years. In the paddock with her were Fionnuala, a daughter of Finola of Leam, Frosty Morning a grand-daughter of Winter Roche and Loobeen Meg a daughter of Thunderbolt. Jane does not have her own stallion but is happy to travel her mares, as far as Connemara if need be, to find the right stallion for them. Recently Jane has joined the Council of the Connemara Pony Breeders Society and is one of their appointed inspectors.

[59] S.I.M. 206-209
[60] S.I.M. 180-181 + O.O.M. 42-44
[61] S.I.M. 203-204
[62] O.O.M. 275-277
[63] O.O.M. 286-287

She enjoys her ponies and her visits to the West. Her farming sons are kindly tolerant of her pony breeding, which is of consequence and not a mere pastime. It is sad when breeders die and there is no one in the family to maintain an established reliable female line. Patrick Coyne of Ardnagreevagh (Loobeen) is one of them and Susan de Vesci another. Jane has made sure these two lines, at least, remain in trusty hands.

The Contributors gallery is full and while there must be others whose portraits would hang easily on its walls, they are people I have not had the pleasure of knowing.

THE PEOPLES GALLERY

The time has come to open the door to The Peoples Gallery which houses portraits of those born and bred in Connemara, those who have lived and worked beside the pony. The gallery walls are long and lined with faces, old and new, of those who have endured the uncompromising nature of the land which they call their own. Faces of those I am lucky enough to call my friends.

Henry Whyte and his family live in a large house at a busy two way junction into Galway. Traffic races by Arus Mhuire, Knocknacarra, which was once a busy B & B visited and re-visited by those who savoured the Whyte hospitality. Now that Henry's wife is no longer able to continue the trade which she enjoyed, she welcomes passers by like myself. You are greeted by a warm, homely atmosphere, with the younger generation of the Whyte family never far away. Henry's wife is often called upon to baby-sit for the young marrieds living in Galway. Bernadette, a child care nurse, has recently returned from three months work in a Rumanian orphanage, and Henry Whyte (Junior) is an active member of the C.P.B.S. Council.

Behind this busy household lies an equally busy yard, which is presided over by the stallion Ardravinna Andy, on whose stable door there is a mighty padlock. When you are introduced to Andy he is invited to take a drink. His head emerges through a side window and a hose pipe attached to the tap is handed to him. He knows exactly how to use it. Andy was bred in Co. Cork and is a strong grey pony who leaves his progeny with a good shoulder and length of rein. Breeders use him well, and travel from the outer reaches of Galway to do so. He is usually well represented by youngstock at the Barna inspection point in the spring.

Henry rents a useful field from the Bishop of Galway. This is the other side of the road from his house and in it grazes Windy Pier, a lovely old fashioned Rebel Wind mare, with her progeny by Andy. The Peoples Gallery contains portraits of many men such as Henry. They are the grains of grit, who together form the rock on which the Connemara pony depends.

The first time I knocked on the front door at Cregg it was in the hope of meeting Cregg Lassie. I was too late, for Lassie had died the previous year. Jim Walsh invited me inside and immediately my eyes were caught by a frame hanging on the wall containing photos of Lassie surrounded by the rosettes she had won at the Oughterard and Clifden shows. To find such a memorial in a Connemara home was unusual.

I have never driven through Oughterard since that day without calling at Cregg. There are no ponies and there is no more for me to learn about Cregg Lassie. I call to see Jim and his sister Nora who are my good friends. There is invariably a kettle and a pan of potatoes boiling on the turf stove and it is always my pleasure to join them at their table. Jim's sight is failing but he continues to cut and bring home his turf and to enjoy a days fishing. Sadly, he is denied one of his favourite pastimes, that of reading, but I do not detect a hint of self-pity. When we visit Jim enjoys a chat about two of his favourite topics, boxing and English politics. The latest news of Henry Cooper, Muhammed Ali and Mrs Thatcher is discussed with enthusiasm. To meet Lassie was an early objective in my research. Her memory was and still is a very clear one for Jim, and I feel I know her through him. She regularly appears in back pedigrees and in my overseas post. I am glad it was she, together with Jim and Little Joe, whom I chose for the dust-wrapper of my first book.

As I drive away from Cregg these days I wonder if time hangs heavy on Jim's hands, now he is no longer so active, cannot read and rarely leaves Cregg? I fancy not. As he sits beside his turf stove his long memory surely feeds his meditative hours...: memories of the mayfly season, the salmon which rises with it and - of course- Lassie and her many years of companionship. And maybe other deeper thoughts I know nothing of. Long may I receive an answer when I knock at Creggs front door.

Canrower is less than a mile down the road from Cregg. It hardly seems possible Mark is no longer there. And indeed that Bridge Boy's stable is no longer occupied. Paddy Geogeghans family have grown up and are mostly away from home these days. The little one who cut her teeth on my business card in 1964 is an Aer Lingus Hostess!! Ponies are no longer an important part of life at Canrower.

Mark Geogeghans reputation for entertaining the ladies was well known, but unlike his good friend Jack Bolger, he was not a great conversationalist. It was hard work to stir his old memory, but his love of the ponies and the ladies remained with him till the end of his days. One lovely autumn afternoon he took Jo and myself for a drive by Lough Corrib. We did not feel very comfortable, or indeed safe, travelling in a side car for the first time. The more we protested the more Mark cracked his whip and the faster we travelled. On our way home, one year, we happened on Mark resting by the river bridge on the approach road to

Oughterard. We stopped to say good-bye. Nothing would satisfy him but that we join him in the Oughterard House Hotel (Sweeneys) for a farewell drink. It was Sunday midday and the lounge was bursting with smartly dressed gentle folk. We - and Mark - were in our every day clothes. It made no difference to his friends behind the bar and we shared a 'Paddy'* among the gentry. What happy memories these remain.

The Connemara pony and the Geogeghans of Oughterard are synonymous. Mark and his father ran a large herd of Connemara ponies long before the days of the stud book and registration. His name as owner of Black Paddy[64] No. 8 is on the second page of Volume I. It would be sad if the seventy year association were broken. I have met Mark's son, Jack, at the Leam inspection point on two occasions recently. His features bear a close resemblance to those of his father. He told me how, as a lad, he worked a gelding son of May Boy[65] to transport the five hundred and fifty poles required to bring electricity from Oughterard to Clifden. It is not easy to explain to a man like Jack the rules of 1992 for registering an every day Connemara gelding.

James Lee of Moycullen was buying Connemara ponies by the end of 1950 but he did not register home bred stock carrying his Moy prefix until some ten years later. A very large number of ponies passed through Jim's hands and I believe you could walk through a field of thirty or more and he would be tested to identify them for you. They were always on offer, apart that is, from one notable exception, Windy Cove, a mare who repaid the family loyalty to her over and again. The last son, bred in her twenty-fourth year, Moy Hazy Cove, is currently a Society owned stallion. I regret I never met Jim. I called more than once but never found him at home. After his death his sons, Johnny and Michael, continued to keep ponies on a reduced scale. My friendship with Johnny was all too short. When he died in 1990, still a young man, it was a grievous blow to his family and the Connemara Society to whom he gave so much of his time. On Johnny's death Michael was elected to the Society Council and retains an interest in the ponies.

Jimmy Cannavan is another Moycullen man. If you visit his Fernville Stud today you will find a new square house and a well built barn surrounded by land grazed by ponies and cattle, and mares visiting the stallion. For many years Jimmy had his own Sire, Jimmy's Fireball. Today he cares for the Society Patriarch, Abbeyleix Owen. Between 1979 and 1986 he stood Garnet Irwin's small thoroughbred sire Speck, who in this period was approved for use on a limited number of Connemara mares. Jimmy has a teenage son John who shares his love of the ponies and who has his own brood mare and youngstock. It is John who gives Owen his daily exercise, walking him in hand through the surrounding country

* Paddy - Irish Whiskey

64 S.I.M. 163

65 S.I.M. 193-194

lanes. Jimmy is a good feeder and stableman whose stock do him credit. He is an energetic Council member and inspector and has recently judged at Clifden, Dublin and in Germany. And perhaps the nicest thing I can say about Jimmy is, if you want to be cheered up, go and see Jimmy. He is the best advertisement for cheerfulness I know.

Jack Bolger[66] opened a Pandora's Box for me in the 1970s. His was another door on which I knocked in 1976 and I returned again and again with my not very reliable tape recorder in my hand. I am unlikely to forget Jack's contribution to my research for Shrouded in Mist. I doubt anyone handled a greater number of the best Connemara ponies in the early stud books than Jack. His memory was very sure and his descriptions very graphic. By the time I met him Jack was an old man in retirement and Teresa, his wife, was more or less confined to her bedroom. His son Willie, his wife and their family were caring for them and running the P.O. and general store next door.

Jack, unusually, was born an Irish Protestant and Teresa an American Catholic. They reared a large family and were happily growing old together in the village of Cashel. When Jack's memory failed Teresa often filled the gaps. Teresa did not enjoy good health and we did not meet her until one day she rose from her bed and joined us in the front room. She was smartly dressed with beads at her throat and a cigarette in a holder. She and Jack charmed us with their individuality and natural openness.

I experienced a genuine sadness and loss when Jack died before the publication of Shrouded in Mist and was deeply moved to be shown into Teresa's bedroom when I called to give her a copy. She sat on the edge of her bed with me beside her and we turned the pages of the book reminiscing about Jack as we did so. I could see it was very precious to her. As I left she asked me to pray for her and to send her a card at Christmas.

Since that time Teresa has died and also, sadly, their eldest son Willie. Many of those who visit Connemara today have met a younger son, Bobby Bolger. Bobby is a chip off the old block and up to his eyes in ponies. He has built, or three parts built, a splendid mansion at Derrygimla where he, his wife and family of eleven live. Bobby is an enthusiastic, busy member of the Connemara pony scene in the 1990s.

Padraig Hynes is a man with a difference. I did not have to find him. He found me. We have shared our mutual love of the ponies many times since that first meeting in 1978. Ponies are at the very heart of Padraig's life and his knowledge of them has matured as each year has passed. By dint of hard work and the ambition to breed good ponies and produce them to the best of his ability, Padraig has set a standard for the Canal pony stud. Canal Stage - sometime 'stage' for the

[66] S.I.M. 108 + 116

Bianconni Car[67] - has now become the stage for overseas visitors interested in the Connemara pony.

Padraig runs a flock of 600 breeding ewes on the Twelve Bens which tower behind the Hynes homestead, lying in a sheltered hollow at their foot. He cannot count good grazing as one of his blessings and takes extra grass keep near Clifden and as far off as Mayo for his ponies and to fatten his lambs. Not content with all the hard work this involves, Padraig patrols the local fisheries at night for the extra income it provides. Padraig's wife, Mary, is equally hard working. She nurses full time at the Clifden hospital and helps with the stock when she is off-duty. His mother, Eileen, is quietly supportive, answering the phone and making visitors welcome when there is no one else at home to do so. The warmth of her heart can be judged by her acceptance of 'Josie' as part of the family. A single lonely man, he arrived on the doorstep one day with two suitcases. He moved into the Hynes loft where he remained until his last illness. Josie was a willing, faithful, pair of hands about the place and even I, a casual visitor, miss his quiet presence.

In 1992 I spent a night in Padraig and Mary's new home, a splendid long, low bungalow with an unequalled view out to Recess. They have worked hard for this and deserve all the pleasure and comfort it surely provides.

Padraig's mare Rambling Home was replaced by Village Laura[68] and her half sister Village Colleen and they have recently been joined by Sliabh na mBan Cailin. Laura was champion Connemara at Dublin three consecutive years, 1985, 86 and 87. This unique achievement was acknowledged by the R.D.S. with the presentation of a special memento. An open cheque would not buy Laura and she is sure to breed good ponies at Canal Stage into her old age.

Padraig Hynes comrade in arms is Henry O'Toole of Clifden, arguably the luckiest man in Connemara. I say this for he is the owner of Village Star[69] whom he bought as a doubtful breeder but who is now recognised as one of, if not THE, best brood mare in Connemara. She has bred Henry one Champion after another and has made many another man envious.

Henry has taken the place of the man known as "the bull with the bowler hat"* and provides a valuable 'Bull for Hire' service for the benefit of Connemara farmers. This means he is constantly on call and travelling one of his handsome Charolais gentleman from one end of Connemara to the other. I suspect they take preference over the ponies. Henry is a true stockman and has no interest in the family Bar, O'Tooles of Clifden. He is another of the worlds workers and, thanks to Village Star, is likely to continue breeding good ponies for many years to come.

Joe Gorham is not a born and bred Clifden man but this is where he and his wife live and have raised their family. Today Joe is a valued member of the C.P.B.S.

[67] S.I.M. 57 [68] O.O.M. 43 [69] O.O.M. 76-77 * Artificial Insemination

Council and Inspection Committee and it is his task to ensure the Clifden show grounds are in prime condition for the third Thursday in August every year. I'm not certain from where his interest in ponies came but I have a suspicion his purchase of Kingstown from John Folan in 1961 may be the link. Many good ponies have passed through his hands since that day. Because he does not use the prefix Kingstown on every pony he breeds (his 1991 Champion Rocky did not carry it) it is not possible to bring them all to mind. His grand brood mare Silver Fort never lets him down and produces a good foal to a different sire each year. Joe's eldest son, Jo Jo, is married and lives and works in Co Mayo but he always manages to be in Clifden on the day that it really matters. Jo Jo is also a Council member and has put many hours hard work into the production of the last two volumes of the stud book. Joe's younger son, Malachy, has the family love of the ponies and is already an expert handler. The name of Gorham is unlikely to disappear from the record books.

Michael Conroy's name appears on the 1941 Society stallion list with Clough Rebel[70] in his care. His son Josie does not remember a year since then without a stallion at Bunowen. I knew Michael in his old age by which time he was hard of hearing and not very active. I will always remember his unique way of calling up his stallion. It was no more than a soft burrr between his pursed lips but it never failed to bring the 'hobbled' pony slowly across the bog in answer to his call. And what is more there was only a kind word to greet him, no carrot or sugar lump! I have spent many happy hours in the Conroy homestead, while Josie, in his own very colourful language has recalled the days of Dun Lorenzo - his personal favourite - Carna Dun, the Arab - as he always referred to Clonkeehan Auratum - and more recently Killyreagh Kim, Abbeyleix Owen and Smokey Duncan.

Josie is a hard working man with a large family to support. His faithfulness to the Connemara pony through both the good times and the not so good times, deserves the grateful thanks of all those interested in the well being of the breed.

Two loyal supporters of Josie and his stallions live but a short distance from Bunowen at Lehid. They are Paddy King and Paddy Conneely and his brother, who are both good neighbours and friends. The Village and Lehid ponies have been bred here for the past fifty years or more and are household names for those who know anything about breeding Connemara ponies. Seventeen year old Village Grey and twenty year old Lehid Lovely Princess continue the family tradition and their stock is always in demand. Paddy King's strong weather-beaten face lightens as he discusses ponies old and new with his many overseas visitors. Paddy Conneely continues to haul seaweed from the shore and to farm the land the way he always has. Last time I visited, he was burning gorse, to act as a blow lamp for fusing the plastic pipe he was installing to carry water from one source to another.

[70] S.I.M. 154

Lehid without Connemara ponies would not be the same. There are no young Kings or Conneelys to follow the 'Paddys' and this is a cloud on the Connemara landscape.

Festy King and his son Festy, own one of the few Carna Dun mares still alive in this area, Power Station. My first visit, on a wild and wet night in 1992, was rewarded by a welcome dram and a conversation of days past which soon dispelled the chill outside. Power Station was a venerable twenty-six years old and showed her age in spite of the obvious care young Festy was giving her. And why not? She had done most things in her day; raced on the strands; given rides at Errislannan; pulled the plough and provided her fair share of foals. Festy has a daughter by Mervyn Kingsmill to take her mother's place. Father Festy's blue eyes twinkled as he told me he bought Truska Brownie for £25.00 in 1959 and how she did not owe them much by the end of her life. She was a small Dun Lorenzo mare who bred seventeen foals, one of whom was Power Station. Today young Festy uses Bunowen Castle as his prefix. You can see the ruins of the Castle from the Kings land. If you want to find them take the left hand fork beside the Ballyconneely Church towards the grave yard. Their solid block built house, the first in the Parish to be built this way, lies below the road to the right, well hidden from prying eyes.

Thomas Burke is a man of few words but happily he has always spared some of them for me. I'm unlikely to forget his courtesy in staying at home to meet me on a fine and perfect turf-cutting day. It was the occasion on which he unearthed the photo of Doon Lass for Out of the Mist. Doonreaughn, Riachains Fort, stands above his homestead and this is where he worked and bred his ponies.

Today it is his neighbour Stephen Keaney who runs the ponies on the exposed Cashel hill. Stephen inherited the Winnie[71] family from his uncle, Thomas King. Winnie the Pooh by Doon Paddy is a grand-daughter of Winnie and she and her daughter are regular breeders. Stephen's ponies do not have many comforts for they live on the hill as Connemara ponies are bred to do. Thomas King was a keen supporter of the early shows but they hold no interest for Stephen. I have a rusty old bit, given to me by him, which once belonged to Winnie. Maybe she wore it when she won her class at Carna show in 1941? In 1992 Stephen had two fillies for registration. One of them he sold to Maria Levinge. I was told, firmly, the other was not for sale for fear her dam Winnie the Pooh might not breed him another filly. It is obvious Stephen intends to preserve this female family, living as they do on the Cashel hill.

It gives me great pleasure to hang the portrait of Michael King who lives at Derryeighter on the Errislannan Peninsular. Michael has been neither stallion cus-

[71] S.I.M. 203-204

todian nor successful exhibitor at the Clifden show. He no longer requires a pony to work the land, but for Michael it is a time honoured custom to have a mare on the holding. If she produces a filly foal she may show him some profit. If it is a colt almost certainly none. A visit to Clifden show may encourage a sale and he is not too concerned whether or not the judge recognises his pony. It is after all an excuse for a day out with the family and Mrs. King makes an entry or two in the produce section. No one was more surprised than Michael when Fudge IV[72] brought fame and English visitors to his front door in 1972. Stephen and Michael are typical of many small farmers across the region. They represent the back-bone of the breed in its native habitat. We trust there will always be men like them, men to whom the presence of a pony is an accepted part of life.

Patrick Conneely's portrait does not hang in this gallery but I am tempted to diversify to share some happy memories of my visit to Doon in 1977. The neck of Streamstown is narrow enough but as it forges its way inland it widens into a bay of great beauty and tranquillity protected, as it is, by high ground either side of it. It is not surprising the O'Flaherties chose a rocky bluff overlooking the narrowest stretch of this channel on which to build a castle. Nowadays there are no more than a few stones to show us where Doon Castle was erected in 1500. More recently, in the eighteenth century, this bay provided ideal smuggling grounds and as a result, the area of Claddaghduff became one of the most prosperous in Connemara for a time.

Patrick Conneely lives near the ruins of Doon Castle and it was his mare, Doon Lily, who encouraged my first visit to this area in 1977. Molly Bawn, who worked the land for the Conneelys, was registered number 38 in Volume 1 of the stud book and her daughter, Molly Bawn II, born in 1924 number 406 was almost certainly by John Quirke.[73] It was she who provided the Conneelys with Doon Lily, by Noble Star,[74] in 1942. She was the first pony to travel to Dublin show from the Claddaghduff region. She won her class as a ten year old in 1952 and the following year her colt foal to Carna Dun, Doon Paddy, was born. Doon Lily died on St. Stephen's night in her thirtieth year. Claddaghduff has always been a centre of pony breeding and the Conneelys must surely take some credit for this. A further happy memory of this region is my first glimpse of Thunderbolt in 1981, as he walked towards me with the Streamstown Bay behind him.

In those early times it was Brendan Burke of Patches who travelled the Society stallions. I was too late to meet Brendan but Patrick directed me to Mikey King of Cushatrough, stallion man from 1959 to 1988. This was an uninterrupted thirty years of loyal service, a record which can be matched only by the Conroys of Bunowen. During those thirty years Mikey handled as many as twelve different

[72] O.O.M. 85-91 [73] S.I.M. 175 [74] S.I.M. 163-164

sires and many of them were not the fashionable ones. His weekly route read; Letterfrack and Tully on Wednesdays; Cashel and Roundstone on Fridays (staying Roundstone overnight) Clifden and Ballyconneely on Saturdays and Fair days. I was once told Mikey knew "how to use the bucket". It is certain sure his stallions could not have undertaken this work load without the help of the bucket!! He knew how to keep a pony fit and happy. I am glad Doon Paddy returned to within a mile of his birth place to spend 1970 and 1971 at Cushatrough. On the first occasion we visited Mikey he appeared ill at ease with all my questions. A game of darts with Jo and a 'jar' in the Claddaghduff bar soon reassured him and we have shared many a good crack since. If you visit his homestead today you will see the Waterford Crystal brought home by Thunderbolt and Mervyn Kingsmill, a happy reminder of their success at Clifden. Claddaghduff is the poorer since Mikey relinquished his task. There is no longer a stallion living beside the road overlooking the peaceful Streamstown Bay.

Petie Molloy and his family are Mikey's nearest neighbours and they will ensure good ponies continue to be bred in this district. They live up a steep bumpy old track and I fear for the safety of the under carriage of my car as I travel up and back down again, but confess, I do not have the courage to make the ascent on foot! It is here you will find the Dooneen ponies, those to be reckoned with in the Clifden ring today. Petie founded his herd on two mares Dooneen Grey and Sukie. He is never in a hurry to sell his youngstock, being a wise and patient man who would rather wait and see how they mature before putting a price on them. He is lucky enough to have some sweet grass in small enclosures overlooking the bay where his mares and foals graze in safety. I am always impressed at the way Petie and his son work in perfect harmony. They appear to have a rare understanding where one invariably knows what the other has in mind with, apparently, no word passing between them. Both are strong men with a gentle approach to the handling of their ponies. And young Peter has been known to take a spring holiday in the sunshine of Majorca before returning happily to the farming chores at Claddaghduff.

I hesitated before hanging Graham's portrait. I was not certain in which gallery it belonged. I hope he would have felt more at home in the chosen second one, among his neighbours whom he helped so often and so generously. The greatest tribute I can pay Graham Tulloch is to say 'he was a perfect gentleman'. Shanbolard was the family home but his love was not for the house and its contents but rather for the land and the stock he ran on it, a love inherited I'm sure, from his grandmother Kate Mary France Lushington Tulloch. After training as an engineer Graham spent some years working with a mining company in the Caribbean. At the outbreak of World War II he returned home to join the Royal Air Force. He was shot down and held as a prisoner of war in Germany for four years. When he returned to Connemara at the end of the war he was welcomed by a series of bonfires which were lit between Recess and Moyard. The people were happy at the safe return of their local hero. Before long he left Shanbolard again, this time for Nigeria, where he spent further time in the mining industry. It was not until the mid 1950s that Graham became a permanent resident in Connemara. Somehow I cannot imagine him anywhere else. Graham had a sure eye for a horse or pony and was soon a valuable member of the C.P.B.S. Council and inspection team. He was a respected Judge and produced many good ponies from his Ocean Connemara Stud. I doubt anyone asked Graham a favour and was refused. The last time I met him was a mere three months before he died in July 1990. I called at Shanbolard one evening and found him about to cover a mare with Loobeen Larry.[75] We stood in the yard and chatted for half an hour or so. I detected a certain weariness in his person. Friends tell me he would have been happy to lay aside the demands of farming and indulge in his favourite hobby of National Hunt Racing. Shanbolard no longer belongs to the Lushington Tullochs but it has been restored and is a family home once more.

Graham would be happy to see his niece, Nicola Musgrave, taking such an active part in the life of the Connemara Society and caring for his old favourite Loobeen Larry so well. Nicola lives with her husband Hugh and their children at Cleggan Farm, where they breed ponies under the Annilaun prefix.

Screebe has long been the home of the Griffins. Colman and his distinctive Windy[76] ponies drew me there again and again. Colman was a dour man who rarely smiled but in the years I knew him he did not enjoy good health. I often found his deep, low Irish tones difficult to comprehend but we grew to understand each other. From the beginning Colman encouraged my research by taking me to meet those he thought might help. As the years went by with no sign of a book, he chided me with the comment "We won't be able to lift it when it is finished". The day I arrived in triumph, with a copy in my hand, I found Colman lying fully clothed on top of his bed. It was one of his bad days. I believe my visit and the book helped to lighten it a little.

Colman was a skilled builder who helped the Berridges restore Screebe Lodge and who built himself a comfortable new bungalow. He was devoted to "Miss Lal" and missed her regular visits when the Lodge was sold. The ponies were a central part of Colman's life. The extra volume of traffic and the loss of a pony killed by the roadside, became an increasing anxiety for him. He lived a further seven years after my visit with Shrouded in Mist, slowly going downhill, but always pleased when I called for a chat. During his lifetime it seemed none of the family were interested in the ponies. How pleased I am to know his son, Colman,

[75] O.O.M. 275-277

[76] S.I.M. 218-219

has retained two mares. May the Windy ponies always be seen grazing the boggy moorland surrounding the Screebe loughs as a sure tribute to their loyal breeder Colman Griffin.

Paddy Folan was reared by his grand-parents at Knockadav, Camus. Their homestead lay way back off the road and in order to reach it you had to ford more than one sizeable stretch of water and the bog through which it flowed (Camus Oughter - the winding inlet). It was not an easy place in which to rear a family or to hold stallions. In spite of this, Paddy was a faithful stallion custodian for sixteen years. The area of Knockadav had little to offer stock but his stallions never failed to do him credit. Breeders knew they could rely on Paddy to visit when he said he would and there was no shortage of foals in his district the following Spring. For the benefit of his family, Paddy left the homestead of Knockadav in 1986 and now lives in greater ease at Rosmuck. His love of the ponies remains ever bright. He has two nice young fillies, descendants of his original mare Irish Harp. He has only a rough patch on which to hold them at Rosmuck with his own better grass-keep away off at Lettermullan. Paddy makes light of the difficulties, and like many another of his generation would not be without a pony.

Tommy McDonagh continues to represent the brigade of the 'hardy originals' and is without doubt, an institution in his home Parish of Roundstone. It was here he ran the corner butchers shop from 1955 until he retired. Tommy, one of ten children, grew up in Derrylea, Clifden and moved to the Roundstone area when he married a girl from Inishee, the Island you can see from the harbours edge. He first registered ponies in Volume IV of the stud book and it would take a good man to calculate how many he has handled since then. He was a great supporter of the Flaps in his youth, though I believe he never rode in them himself. His sporting skills were as Champion Miler (on his feet) and as an unbeatable competitor on a bicycle, when bicycle races were a popular sport in Connemara.

On the first occasion I was to meet Tommy, I asked directions to his butchers shop. "You can't miss it", I was told, "all the Cups he won at Clifden yesterday are in the window". Sure enough, not a lamb chop to be seen, but the Archbishops, the Clonkeehan and the Jan Harold Championship trophies stood proudly on the shelf. Roundstone River, the worthy winner, was stabled behind the shop. The Clifden Champions Cushatrough Lass[77] (1956) and Dairina (1957) were produced by Tommy, and the stallion Gael Linn was among his exports to the U.S.A. He handled the Society stallion Noble Star in the 1930s. I have a suspicion Tommy's greatest pleasures in life were having a deal and discovering a good pony on an obscure mountain top. For many years he was a Society Council member and is now a respected Vice President. We hope he will be around to cast his experienced eye over the ponies of his friends and neighbours for many a year yet.

[77] S.I.M. 238 + O.O.M. 58-59

Willie Diamond is another man of much the same era. He owns Diamonds Bar at Tully Cross but I do not know whether he has a similar claim to Diamonds Hill!! Because he has always proved elusive I do not know Willie very well, but Diamonds have long been connected with the pony. His father, Michael, handled Society stallions between 1928 and 1937. Willie himself bred and owned Kimble and kept the stallions Roundstone Oscar and Davy D for some years. He still has a mare or two about and we expect to see him at Clifden in August every year.

The signpost to Noel Mannion's home has always been a white circle on the wall opposite. Now it is a 'dish' on the wall of his homestead. Noel believes in keeping up with the times! When I called in 1992 he had just returned from a day on the hill, working with his sheep which, he tells me, have taken the place of the ponies as his number one interest. Noel hit the headlines a few years back with his prize-winning mare Grey Rock Star, and is a reliable source of information concerning pony breeding in his area. I discovered there are many breeders between Ballyconneely and Roundstone whom I have never met, at Lettershask, Forglass, Emlaghmore and Douhulla. When I left, Noel was on his way to Saturday evening Mass "I'm not a religious man" he told me "but I would not miss Sunday Mass"*. I am aware of the strong bond between the Church and the people. To attend weekly Mass is 'belonging' to the Community and has always been at the heart of the culture of the West.

The De Courcys of Errisbeg have a long association with the Connemara pony. John was a stallion man in the earliest years of the Society. Today it is his son James who owns a stallion Slisneoir, son of the popular Cuchulainn.[78] James has built a useful sand ring below his stable yard and this is where he trains his ponies to jump. He has found show jumping an attractive pastime and travels many miles to South Galway to compete with his ponies. The Roundstone show has become a popular annual fixture in the Connemara calendar and it was James who was responsible for its introduction in the 1970s.

I have to confess I have never fully explored Carraroe. It is another of the wondrous Connemara Peninsulars and lies between the bays of Casla and Greatmans. It is an active pony breeding area and the Derrynea spring inspection point is well patronised by breeders from Carraroe. There are two men living there whose portrait must be hung and they are the brothers O'Malley from Derrarthabeg. They have cared for a Society stallion since 1979 and Seafield Silver Rogue, not the easiest of ponies to handle, since 1988. I was pleased and surprised to discover Bartley had travelled all the way into Clifden for the launch of Out of the Mist. I met him the following morning carrying his copy as if it were a treasure. I hope he will want to own a copy of these Reflections.

*Saturday evening Mass is the first Mass of Sunday

[78] O.O.M. 275

Cnoc Mordáin casts its long, dark, shadow over Loughaneera, a small townland on the road between Screebe and Kilkerrin. In the earliest days of the stud book it was a centre of pony breeding with Thomas Cloherty and Pat O'Sullivan (John) showing others the way. Thomas owned Cashel Lass registered No. 11 in the stud book. She won her class at the Society Show on four occasions between 1933 and 1939. When I visited Mark Cloherty in 1992 a photograph of Cashel Lass with his Mother at her head stood proudly on the sideboard. There were no registered progeny from Lass in the stud book but Mark tells me Cnocmordain Lass was one of hers and that another went to Pat John. Who is to say this was not Lougheera[79] herself? Cnocmordain Lass has two registered daughters one of whom was Winter Mollie.*

The mountain of Cnoc Mordáin is named after a wizard by the name of Mordan. Legend has it, that Mordans shepherd owned a fierce hound who leapt at St. Ciaran, as he passed by on his way to the Aran Islands. As the Saint ducked the hound flew over his head and fell into (or created by its impact) the lake of the 'shepherd's hound'!

An Anacair, the rough stony place, below Cnoc Mordáin is where Mark runs his mare and foal. It is a very unwelcoming grazing area with little grass for the ponies to forage, from among the rocks and boulders. I did my level best to follow Mark's strong sure footsteps from one boulder to the next but was grateful, when he saw I was flagging and told me to wait. I sat on a boulder and marvelled as I watched his return with Winning Ways and her eight day old foal by Moy Hazy Cove. They seemed to make nothing of the rough terrain. Mark has had the Winning Ways family since the 1960s. They have no relationship to Cashel Lass. Mark is a bachelor and is the last of the Clohertys of Loughaconeera. As I looked at his strong, bronze face and dark hair, I wondered if he came originally from Spanish stock? And maybe Cashel Lass also? This coast line is said to have been one of the areas from which the Armada suffered. A point a little further along is named Cora na gCapall and a wee Island beyond Carriag na gCapall.[80]

I took a photo of Pat O'Sullivan's closed and empty homestead and left Loughaconeera with a sad heart. It seems likely pony breeding here will die with Mark.

Towards the end of the gallery on the left hand side, are portraits of three men who do not live the life of the Connemara smallholder. They derive their living from other sources on the fringes of Connemara in Mayo and Spiddal. This in no way devalues their contribution to the breed and the Society, but the story behind each portrait is a little different.

John Luskin's twin business, a timber yard and a passenger launch, thrives in the historic Mayo town of Cong. It is situated on the ithsmus of Loughs Mask and

[79] S.I.M. 212 * O.O.M. 118-119 [80] Horse Islands

Corrib where John's launch provides tourists with the opportunity to visit Ashford Castle and to savour the beauties of Lough Corrib. The Cong woodlands are famous for their cock and pheasant and it is here, no doubt, John finds his timber. In the nineteen seventies he had a reputation for producing the winner of the stallion class at Clifden. Macduff, Carrabaun Boy and Milford Hurricane immediately come to mind. John was my co-judge at Clifden in 1985. We had a long damp day in the centre of the ring but soon established a good working relationship. I am pleased to know he retains his interest in the pony and is currently a Council member, where I'm sure his long standing knowledge of the breed is appreciated by some of the younger generation.

Spiddal will always have a special place in my heart because of Michael Clancys generous hospitality on numerous occasions. It is curious to look back to 1978 when I was invited to judge with him at Dublin. We did not know each other in those days but had no difficulty in agreeing how we would place the ponies and, had the weather been kinder, we would have enjoyed each others company. That pleasure was deferred until 1981 when I was living at Errislannan and occasional visits to Galway for research were necessary. The 'open door' hospitality of Michael's Bridge House Hotel often restored me at the end of a tiring day. The same welcome greets me whenever I return to Spiddal and this is something I value greatly.

Michael's Atlantic prefix is probably better known than the man himself. It has spread world-wide and so many good ponies have carried it. I wonder how many? Michael had a habit of combing Connemara before weaning time, leaving his buyers mark on as many as forty foals. He became a shrewd judge of young stock and his contribution to the breed is incalculable. He was a keen show-man, regularly at the top end of the line with his ponies. Once he had achieved the supreme Championship in 1985 with his mare Silver Sparrow[81] it seemed his ambitions were fulfilled. Today when his ponies appear in the ring it is his enthusiastic stableman, Peader, who brings them there. Michael's primary interest now is in National Hunt Racing and breeding thoroughbreds. I was encouraged to find a new young stallion by Mervyn Pookhaun in his yard in 1992. The ponies must still hold a small place in his heart.

Almost directly opposite Bridge House Hotel is John Enda Feeney's butchers shop, behind which is a modern abattoir which fulfils current EEC regulations. The Feeneys buy and kill all that is sold in the shop, which is some compliment to John Enda. His son Sean shares his interest in the ponies and a barn style range of boxes and an indoor school have been built within two miles of the shop. They have a useful sire in Gorteen Boy and use Aran as their prefix. The Feeneys are the oldest family in Spiddal and John Enda's presence on the Council of the Society

[81] O.O.M. 285

must be a welcome one. If you do not know John Enda by sight, look for the man carrying a twig! It is his modest but effective form of protection as he inspects ponies or cattle. "Stand off" it says with its gentle prod. You will not meet John Enda about his business without this powerful weapon in his hand.

One of the many side roads off the main Spiddal to Galway coast road leads to Baile An Dommullain, Ballydonnellan, which I discovered, consists of no more than a cluster of dwellings. This was where the mare, Ballydonnellan Grey,[82] lived and worked and I was anxious to meet her owner, John Conneely (Morgan). When I called in 1990 John was not at home. It was a warm, sunny May afternoon when I returned in 1992. As I walked into the garden I saw the back view of a figure half buried in a large shrub-like hedge. He emerged as I said Hello and together we investigated a blackbirds nest full of chicks in its centre. I talked to John about his ponies Ballydonnellan Grey and her daughter Breidin,[83] and learnt how Grey was won as a prize in a raffle by his daughter. I saw the cart she had worked in and the hill she had lived upon. She had, John told me, died 'on pension' after her breeding days were over. For the Conneelys she had been more or less a gift. What a gift to the world, Germany and Sweden in particular, her progeny have been. I sat in John's garden with a glass and two photograph albums in my hand. Yes, it was a further gift to find old photos carefully preserved. John Conneely had clear blue eyes which clouded as he told me his wife had died three years ago and his son and daughter, both married, lived away from home. His garden was full of rather crude sculpture, moulded by him out of spare mortar, which he told me, it was a shame to waste. The topiary of his garden hedges was more elegant and his artistic leanings implied a latent talent. During the hour or more I spent with John I relived the lives of his two good ponies. I doubt we will ever meet quite their like, nor indeed the like of John himself, again. They are the golden tinted threads running through the carpet weave of days past.

The first whiff of Galway calls me to speed through and beyond the city and to Connemara and all the joys it holds for me. For this reason I have spent all too little time South and East of Galway and this is my only excuse for not having met Murty McGrath before 1990.

Murty lives at Cregmore, Claregalway where he has a substantial farmhouse and an acreage which, I guess, far exceeds that of the average holding in Connemara. It is good land and Murty is surely a good farmer. He has been a long-time supporter of the Connemara pony and his original mare, Smokey, has left him a valuable female line. Murty is also blest with a valuable male line of McGraths with Patrick, Kiernan and Murty (Jnr) all in their twenties. Mrs. Murty provides a great spread on her long table where, I discovered, Anne Hemphill is

[82] O.O.M. 169 +227 + 281 + 283 [83] O.O.M. 227 + 283

very much at home. Anne and Murty have been good neighbours for many years. Murty probably knows as much about Connemara ponies as many will have forgotten and I'm sure his sons benefit from his wisdom.

While I may not have met Murty until recently the Ganty prefix led me to Joe McNamara at Craughwell many years ago. Joe was a great enthusiast who earned his living from farming but whose ponies were a welcome source of additional income. Joe was a family man and they always came first but the ponies were a good second!! My best memory of Joe was his arrival at O'Gorman's launch party for Shrouded in Mist. It was held in the Galway shop after closing time and those who came were greeted by Philip O'Gorman, and a glass of sherry. It had received little pre-publicity but Phil McDermott rang a few local breeders and Joe was one of them. His cheerful presence was very welcome, especially by myself. Phil noticed Joe leaving the shop without a copy of the book. Joe fished in his pockets. There was insufficient there to pay for one. I said to Joe "Please take a copy. It would give me great pleasure for you to have one". He insisted on leaving all his loose change. The following year I was judging at Clifden. Joe walked into the centre of the ring with the balance in his hand. There is, I'm sure, no need to say more about Joe McNamara, except, how much all of us who knew him miss him.

To close the door of the Peoples Gallery is not easy. Its portraits hold so many happy memories. I could have filled another gallery of equal size without difficulty. However, I see the door to Gallery number three is ajar and it is time I walked in.

The life blood of any Society flows from its officials, those who are responsible for the day to day turning of the wheels. The C.P.B.S. has been very well served by Society officers, who are relatively few in number, thanks to the long service most of them have given.

The O'Malley brothers of Derarthabeg with a friend, and the stallions Ocean Minstrel and Abbeyleix Owen in 1980.

Mark Cloherty's mother and Cashel Lass, a regular winner at the Connemara pony show in the 1930s.

THE STEWARDS GALLERY

The first portraits in the small stewards gallery hang side by side. They are of course Michael O'Malley[84] and Bartley O'Sullivan[85] Their dedication as joint secretaries of the Society for the first thirty years of its life provided a very sure foundation. I can tell readers no more about these two men. My regard for them remains as great now as it was when I wrote Shrouded in Mist. This gallery would be incomplete without their portraits.

John Killeen's portrait hangs beside that of Bartley. For three years after Bartley's death the secretarial work was undertaken from the Committee of Agriculture Offices in Galway. In 1963, John Killeen, who was already sixty-six years of age, was appointed secretary. He had recently retired from his post as Supervising Inspector of the Western Region, which covered the seaboard from Donegal to Cork. As a result he was familiar with rural life and was a fluent Irish speaker. At the time of his appointment he was living on Nuns Island, Galway but by the time I knew him he had moved to Ard na Mara, Salthill. It was his good friend Sean Keane, who introduced him to the people and the ponies, with whom he was soon on easy terms. Whatever hour of the day you called, often without an appointment, you would find John Killeen at work behind his desk. His sixteen years as secretary were distinguished by the careful, courteous way he carried out his duties. I doubt there were errors in the stud book, or lack of welcome at his door. And most important of all, he understood the people and the difficulties they sometimes experienced over registrations and the like. He was never ever dismissive of their needs.

Over a period of years he came to trust me and was tolerant and helpful, however often I called on him. He finally allowed me free access to such old records as the Society had, notably the battered stallion file. On my last research visit to Ard na Mara, he left me on my own in the office at lunch time with a sherry glass of whiskey and a bar of fruit and nut chocolate.

John Killeen died in October 1979 and was found sitting at his desk with his pen in his hand. Two days previously I had seen him at Maam Cross Fair taking particulars of transactions. Those who knew him are unlikely to forget his sixteen years loyal stewardship.

The portrait of Phil McDermot hangs beside that of John Killeen. It was she who took his place in the secretarys office. Phil had a long association with the Society. She worked part-time in the Department office in Galway in the 1960-1963 transitional period and continued as part time assistant to Mr. Killeen. She was an obvious choice to take his place in 1979. Phil ran the office from the front

[84] S.I.M, 95-96 + 129

[85] S.I.M. 128-130

room of her house without any help. As a mother and house-wife and an enthusiastic golf player she was carrying a heavy work load. Phil had a chirpy sense of humour and an uncanny memory for stud book numbers and pedigrees. She knew her work and was solely responsible for the stud books and show catalogues, as well as dealing with the daily round of secretarial chores. The Society owes her a debt of gratitude for her eleven years loyal service in the secretarys office.

On the wall opposite the secretaries portraits, hang those of the Society Presidents. I very much doubt the Presidents role in the C.P.B.S. has ever been a nominal one. When the Society was formed in 1923, with Father White as President, it was obvious he played a fully supporting and active role as a founder member. I am not certain whether Father White died or moved out of Connemara but he was President for a very short period, as was his successor, Monsignor McAlpine, Parish Priest of Clifden. In 1937 Father Cunningham, P.P. of Roundstone was elected President. By 1943 he had moved to Clifden and been appointed a Canon. He remained President until his death in 1952. Canon Constantine Cunningham was a much loved and respected P.P. and we can be sure his long service to the C.P.B.S. was a valuable one.

Lord Killanin's name is included as a Vice President in 1947. It was a break with tradition when he was invited to act as President in 1953, thus taking the place of the Clergy who had held office for thirty uninterrupted years.

It may be surprising that a man of his standing in the horse world should have considered the Connemara Society worthy of so much of his time and expertise. A great tribute to the Connemara pony surely? Lord Killanin was a committed Galwayman with a house in Spiddal and in many ways was 'one of the people'. His presence among them was always respected and appreciated.

He was born Michael Morris, great grandson of the first High Sheriff of Galway, and inherited the peerage on the death of his uncle, the second Lord Killanin, when he was only fourteen years old. The Morris association with Spiddal dates from 1686 when a member of the family married a niece of the notorious Roderic O'Flaherty, who owned land in Spiddal at the time of the marriage. Indeed the Morris's were one of the Galway Tribes and in 1486 Richard Morris was a Bailiff of Galway.

Lord Killanin was Chairman of Galway Races, which he would never willingly miss, for sixteen years. He was a member of the Turf Club and Chairman of the Irish Government Commission on Thoroughbred breeding. He continued to act as President of the C.P.B. Society until he took up his appointment as Chairman of the International Olympic Committee in 1973. During his twenty years service as President there were two important land-marks in the life of the Society. The decision to close the stud book in 1963 and the staging of the Interna-

Thirty Year Review

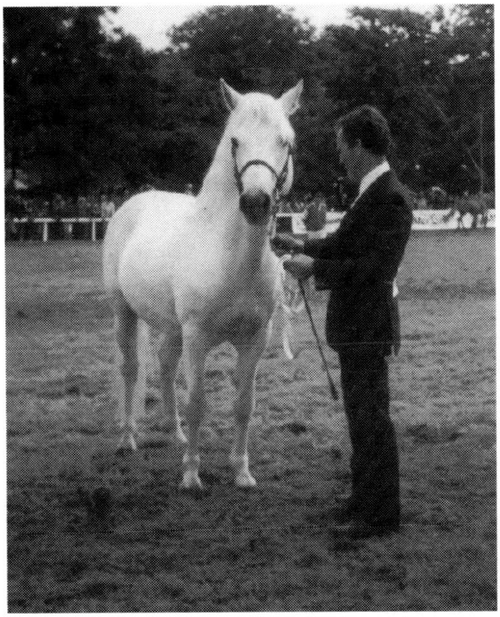

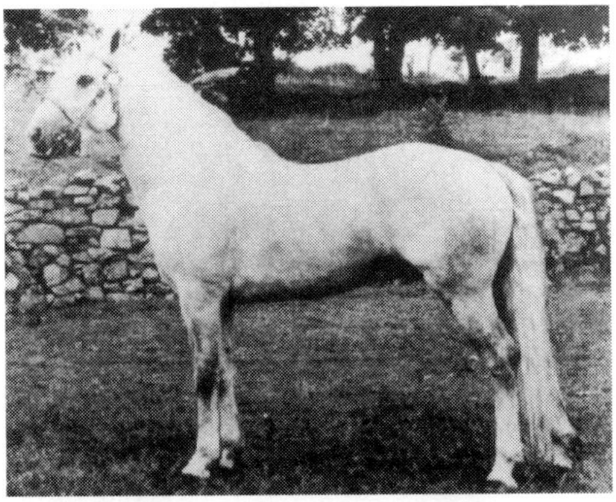

Two of Rebel Wind's sons: Dunlewey Sonny Jim and Tamson Clarke (top L) Murphy Rebel and Phillip McMahon at Clifden in 1990 (above R)

Greaney Rebel, winner of the stallion class at Clifden 1968 (L)

Sarsfield at Cregduff in 1992 (below L)

Killyreagh Kim with Graham Tulloch and Colonel Crighton at Shanbolard 1968 (below R)

Thirty Year Review

Calla Boy (above L) at Ganty in 1991

Tulira Finn MacCool Champion stallion, Clifden 1991 (above R)

MacDara (R) in his 21st year at the Tulira Parade. 1970 (R)

Ben Lettery (below L) on Lambay in 1975

The Fugitive (R)

Thirty Year Review

Mervyn Kingsmill (above L) as a three year old at Clifden in 1983 with Petie Molloy at his head.

Mervyn Blue Charm (above R)

Marble (R) ridden by John Brennan (Jnr), Clifden, 1973

Strongbow in Germany (below L)

The Admiral in his 17th year with Timothy Carey, 1977 (below R)

Thirty Year Review

Ardravinna Andy (above L) takes a drink

Doon Paddy with Joe Macnamara (above R)

Tully Grey (centre L) ridden by Evelyn Harrold in 1964.

Errigal Prince (R)

Thunderbolt (below L) and Mikey King

Ormond Oliver in his 17th year, 1992 (below R)

Thirty Year Review

Yarraman Park Henry in Australia, 1976 (top L)

Padraig Hynes with Sliabh na mBan Cailin, Champion mare Clifden 1992 (above R)

Ballydonagh Deidre, Supreme Champion at Heteren, 1992 (L)

Pat Folan of Callowfeenish (R)

Mary Ward and her two year old filly, Pigeon (below L)

Lehid Lady Belle (below R)

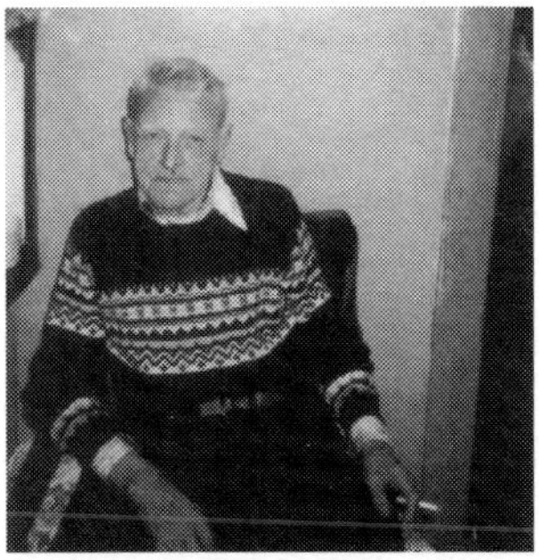

Thirty Year Review

Two of Cregg Lassie's sons.

Little Joe (top L) with Sean Stagg in 1991.

Cregg Hero in Sweden (top R)

Gregory (above L) ridden by Linda.

Leam Little Lady, in her twentieth year 1993 (above R)

Woodlands Rover (L) bringing home the hay, 1978

Thirty Year Review

Errisbeg Rose (top R) in Oxfordshire and her daughter Errisbeg Dolly (top L) at Aillebrack, Connemara
Camus Harp (centre L) in Normandy and (centre R) 25 year old Abbeyleix Molly at the Coosheen Stud in 1989.
Martin O'Donnell (bottom L) with his mare Swift Sally and her foal by Smokey Duncan and (bottom R) Millie of Abbeyleix in her 23rd year at the Tyan Stud in 1992.

Thirty Year Review

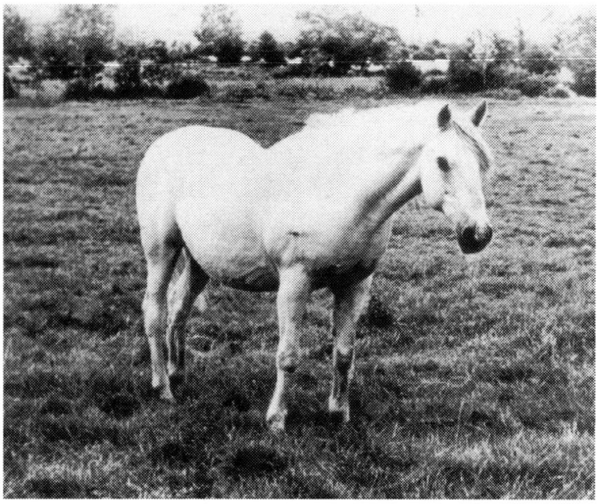

Gloves Misty (top L) with her owner Eddie Madden, Champion at Clifden in 1982.

Charm Bracelet, at Glann in 1984 (top R)

M. Basguinus' French Dressage Champion, Tulira Crackerjack (centre L)

Cregmore Colm, in New South Wales 1981 (above R)

Smokey (L) at Tulira in her old age.

Thirty Year Review

Cailin Ciuin with her first foal 1993 (R)
Wireless Wave at Alnabrone (centre L)
Knockabout Lady at Kylesalia, 1992 (centre R)
Callowfeenish Beauty at Moycullen (below)

Home Lover connection

Loose Ends

Pillagh Grey Pearl at the E.C.P.S. Show 1992 (above L)
Celebration Cream at the E.C.P.S. Show 1979 (R)
Toon Duinhof (R)
The turf carts at the Claddagh early in the nineteenth century (below)

Loose Ends

Dr Nicola Hall and Silver Duke (top L) at the E.C.P.S. Ridden Show 1985

Errigal Prince with Michael Connolly (top R) winners of the 13.2. and under Jumping Class at Ballyglunin Gymkhana 1968, surrounded by his brothers and sisters

Ocean Bambi (L) at Ballinaboy in 1989

Milady de Ravary in 1991 (below R)

Ocean Seames in Germany 1976 (below L)

Loose Ends

Indra Rebel (above L) at the St Lo Haras, 1991.
Ardan (above R) in May 1992, a month before her 30th birthday.
Barna Honey (below L) and Bill Concannon at Clifden Show 1967.
Mark Osborne (below R) and Wychwood Grey Gull in 1991.

tional Conference in 1970. It seems reasonable to assume that his steady hand and leadership steered the Council through these significant stages in the growth of the Society.

It was Tadgh O'Sullivan, son of Bartley, who introduced me to Lord Killanin in 1981. I met him at his Lansdowne road home in Dublin and we talked for an hour. His study was lined with a treasury of antiquarian books and of photographs recording many interesting high points in his life. In September 1983 I asked Lord Killanin if he would write the Foreword to Shrouded in Mist. He kindly agreed, on condition he had an opportunity to read the complete text before doing so. He was not very happy with Part One because of my attempt at a potted social history of Connacht. He wrote suggesting the copy be read by an Irish scholar, possibly an expert on Irish ponies. I had no idea where I would find such a person and Shrouded in Mist had become so personal I could not imagine letting anyone loose with it! I immediately sent Lord Killanin a copy of my Preface followed by Part Two. He accepted both in good faith and wrote the Foreword which appears in Shrouded in Mist. I was sad he remained reluctant to give it his wholehearted blessing. A letter I received later helped to redress the balance. After I left a complimentary copy of the book at his home, his letter of thanks read "Last night I was able to go through it again. The finished work is certainly a great tribute to your industry and love of the West."

Sean Keane acted as a dependable bridge on more than one occasion and was the link with Connemara when the Committee of Agriculture undertook the secretarial work in 1960. He was co-secretary with John Killeen when he first took office and was an important member of the Council for some years. I believe it reflects the regard the Council had for his knowledge of the pony and the region when he was elected President to succeed Lord Killanin in 1973. He could hardly have been an easy man to follow. I cannot say I knew Sean Keane, though I met him more than once. Many of the younger men in Connemara today will tell you how he encouraged them and shared his knowledge with them. I believe the working relationship he and John Killeen enjoyed was similar to that of Michael O'Malley and Bartley O'Sullivan. Sean Keane died in 1978 and thus held the Presidents position for five years only. His service to the Society was far greater than these five years indicate.

I first met Donal Kenny in 1976. He had recently been elected a Council member of the C.P.B.S. having already served four years as an Inspector. He was living in Oranmore and working at the Department of Agriculture office in Galway. I found myself sitting beside Donal at lunch after our Societys mutual discussion concerning COPA.[86] We were soon at ease with each other. Donal told me

[86] COPA was a group of Farmers organisations in the EEC (Pony Breeders considered farmers for EEC purposes)

he was studying for a Masters Degree in Equine Genetics and I told him of my dreams of writing a book on the history of the Connemara pony. He left with a copy of The Chronicle in his hand and I with the knowledge that he would help with my research in any way he could.

In 1978 Donal Kenny was invited to take Sean Keanes place as President of the Society. By this time he was living and working in Dublin and it presented a demanding commitment. When he accepted, I believe it was with the intention of being a 'stop gap', and therefore, of not remaining in office for too long. In the event he was President until his resignation in 1990. For twelve years he gave generously of his spare and holiday time to the interests of the C.P.B. Society. His knowledge and gentle diplomacy were greatly valued by those who knew him. He was a good and helpful friend to myself, and we met in Dublin to discuss my work more than once.

In 1984, on his return from an overseas business trip, Donal sat up half the night in order to read Shrouded in Mist before travelling to Clifden the next day. He, quite rightly, was not prepared to give the book his blessing at the pre-show reception without having read it. I was very pleased when he agreed to write the Foreword to Out of the Mist and appreciated the generous welcome he gave the book.

An extension of my friendship with Donal was a visit to Lesotho to see the Basotho Pony Project first hand. This was something very close to his heart and was sure evidence of his skills in his work for the Ministry. It was a memorable visit for me and I felt privileged to have made it with Donal as my guide.

Towards the end of his Presidency Donal prepared a set of rules for the arrival of the single market in 1993. He worked in close co-operation with the English Society and kept an open mind to our problems and indeed, to those of others throughout the world. His carefully drafted rules have formed the basis for an International Committee for Connemara Pony Societies. They will be a lasting memorial to Donal's generous and thoughtful contribution to the Connemara Pony Breeders Society.

The stewards gallery only contains portraits of those who no longer hold office. I hope space left at the end of each wall will remain vacant for years to come. The continuity of officers has been a hallmark of the seventy year life of the Society.

CHAPTER THREE

THIRTY YEAR REVIEW............1963-1993

The decision to close the Connemara Pony Breeders stud book was taken in December 1963. By this time one hundred and fifty eight stallions had been registered, but only four male lines had survived and stood the test of time. They represented forty years dedicated work by the Society, during which they had sifted the wheat from the chaff, taken the risk of introducing outside blood, and encouraged the people to register their ponies. All this with the single minded objective of restoring and preserving the best attributes and character of the indigenous native pony.

The 1964 Society stallion list contained the names of thirteen sires, who represented the four blood lines. They could hardly be regarded as a pot of gold. I plan to follow the progress of each line through three decades, by including a reference to every sire used by the Society in this period. Occasionally I stray from my remit to include a pony or a personal memory, which I hope adds a little flavour to the overview.

A fifth and valuable native blood line re-surfaced in 1969 and proved a useful addition to the limited gene pool.

I began this review with eager anticipation. No charts, no stud book numbers, a happy ramble among the ponies, many of whom I know so well. In reality, I discovered that to work without the props of charts and pedigrees rendered my task a testing one. None the less, I determined to stay with my original plan and became absorbed by the progress of each line, seeing how they have flowered or wilted or withered in thirty years. A small concession to charts has been made, in order to show the ponies who represented each of the four lines at my starting point in 1964. The four lines are (1) Cannon Ball/Rebel (2) Connemara Boy/Dun Lorenzo (3) Little Heaven/Carna Dun (4) Naseel/Clonkeehan Auratum and (5) Mountain Lad/Tully Lad.

CANNON BALL	REBEL WIND .... 4 yr. old son of INVER REBEL
REBEL LINE (1)	BRIDGE BOY .... 5 yr. old son of TOOREEN ROSS

In 1964 Rebel Wind was the only sire from the number one line to travel the Connemara roads. He was a doughty little character who gave fifteen years sterling service to the breed. We met many of his good sons as we travelled the world in Out of the Mist and those of us lucky enough to own one of his two hundred and eighty-six registered daughters appreciate their worth. The flea-bitten greys are attractive, the bays, and strange to say they often came that colour, rather less so. Happily the colour makes little difference to the end product of a brood mare. Rebel Wind never hit the head-lines in the show ring and it is true to say many of his best progeny have not done so either. He was on the small side and the perfect gentleman, who Paddy Folan remembers as his favourite sire.

Only two of Rebel Wind's sons were chosen to follow their sire onto the Society list. Roundstone Oscar 1973 - 1977 and Murphy Rebel 1976 - 1989. Oscar's dam, Dancing Spanner, was by Little Heaven and was registered without her dam being named. She raced on the Connemara tracks as Gold Wing. Oscar spent five seasons in Connemara before leaving for Canada. He left two unexceptional sons and twenty-seven registered daughters behind him. It was not a great record.

Murphy Rebel, on the other hand, is proving to be a worthy son of his father. He was privately owned by Stephen Heaune of Clifden and did not travel, which limited the number of mares he covered. His dam, Muffy,[87] was that grand old Auratum mare who lived on the sky road and who bred well to a different sire every year of her productive life. In 1990 Murphy was sold to Co. Mayo and from there to Philip McMahon, of Belturbet, Co. Cavan, where he has taken on a 'new look', carrying more condition and showing many of the pleasing traits of the breed. He has since won his class at Galway and stood well up the line at Clifden. It is only since he left Connemara that his progeny have come good, notably the fillies Coral Misty and Moorland Juliet, both winners at Dublin and Clifden. In 1992 I awarded a ridden Mountain and Moorland Championship to a very charming daughter in England, Bramble Lass. I am sure there are more of similar calibre in Connemara. Philip tells me Murphy covers up to eighty mares a season, though only twenty or so are registered Connemara mares. Bobby Bolger has a useful son in Coral Prince and two more are likely to be registered from the 1992 inspections. A son, Catamaran, stands in Oughterard and Ferdia, a three year old son of Wireless Wave, looks a further likely prospect.

Another interesting son of Rebel Wind lives with the Clarke family in Kilkenny where he leads a very full life. Dunlewey Sonny Jim[88] grew up in Co. Donegal on the Guinness stud in Letterkenny. As a four year old he moved to Kilkenny where he teamed up with seven year old Tamson. Together they have attended the North Kilkenny Pony Club thirteen successive years. In 1991 he gave six weeks valuable service to the Camphill Community for handicapped children. In between times he covers an average of five mares each season, hunts with the local pack of foxhounds and takes part in all Pony Club activities. I hope he leaves many as versatile as himself in County Kilkenny.

When I last enquired, twenty-three-year old Dangan Dun*, Rebel Wind's senior son, was alive and well. He was bred at Bunowen from Breath of Spring[89]

[87] S.I.M. 114 [88] O.O.M. 279 * Dangan Dun was put down after this was written [89] S.I.M. 221 O.O.M. 134 + 204 + 276

but has spent his entire adult life at Lower Dangan, Galway, with Paddy Carr. His registered progeny are modest in numbers. Because his son Atlantic Cliff[90] is the sire of Mervyn Kingsmill, Dangan Dun's name will feature in many a back pedigree for years to come.

After the publication of Out of the Mist I was chastised by Dr Reijnders of the Netherlands for including a photo of Atlantic Cliff without the permission of his daughter, Anne Marie, who took the photograph. Cliff, he told me, is no longer fit for stud duties or for jumping, but in Anne Marie's care he leads a happy life and together they enjoy competition in the dressage arena. For some years he was in regular use at an Institute for mentally handicapped children. A further proof, if we needed it, of Rebel Wind's precious gift of intelligent kindliness.

As I reflect, I am puzzled as to how Mervyn Blue Charm, dam of Mervyn Kingsmill, met Atlantic Cliff. Cliff was bred by Michael Clancy but was registered by John Daly. The stallion records indicate he was at Lough Mask House when Blue Charm visited him in 1979. Did he pass through Dublin on his way to Holland that year, or did Charm travel to Mayo? One thing is certain, he gained the approval of Lals' shrewd eye.

Mervyn Kingsmill spent eight busy seasons (1983-1990 inclusive) in Connemara and was one of the few Society stallions to live the twenty-four hour a day open air life, hobbled and grazing happily in one of the small Streamstown Bay enclosures. He did not trot the roads beside Mikey King's bicycle as stallions had in years past. Mikey, with a replacement hip, invested in a single pony trailer for his travels. Kingsmill was soon a popular sire. Many of his progeny were in the ribbons at Clifden, as indeed, he was himself. It was under John Luskin and myself he won the stallion class in 1985. We considered he had a touch of class and sufficient bone and substance to take him to the top. He followed this with 2nd place in 1986, 3rd in 1988 and won the confined class in 1987. By 1989 he was out of favour and unplaced. At the end of the 1990 season he was sold to England. Kingsmills critics felt he left too many indifferent progeny who did not improve with maturity. In his place the Society have two of his sons who were at work in Connemara in 1992, Village Boy and Matchmakers Lad. A lot will be expected and required of them. There are many good Kingsmill females throughout the region and he may be said to have done his job for his homeland. There are two privately owned sons in the district, Streamstown Larry and Golden Discovery. I judged Kingsmill for a second time in England in 1992 and sadly failed to recognise him. He appeared to be missing his native Irish air and did not show himself to advantage. He provides English breeders with a golden opportunity and I hope the journey to Suffolk will not deter them.

[90] O.O.M. 205 + 276

Before we leave Rebel Wind and his Irish dynasty, a word for his finest son, Marble. In his homeland Marble's contribution to the breed was a nominal one. He lived at the top of Taylors Hill overlooking Galway City and his flow of visiting mares was slim. I wonder how different this story might be had he travelled the Connemara roads. Marble's double at Clifden in 1973, when he won the Open Stallion class and the Ridden class, with an equal degree of effervescence and good manners, is a memorable one. His departure to Denmark in 1976 gave him the opportunity to prove himself a good sire, and this he did across the quality mares of the Oxenholm stud. Denmark and its European neighbours gained what Ireland lost. The last remnant of this male line in Ireland is a grandson Wild West Rain who stands in Drogheda.

I wonder how it was that by 1964 a son of Carna Bobby[91] had not presented himself as suitable for inclusion on the Society stallion list. In 1961, after twelve years service, Bobby left Connemara for Gort. Is it possible his contribution as a sire was not recognised until some years later? It is generally agreed that Carna Bobby was weak behind and that this was reflected in some of his progeny. It may not have been considered prudent to retain a son of his for this reason. However, in 1968 a colt by him was purchased from his breeder Joe McNamarara. He was named Ardilaun, (after an island, situated near Cong), and joined the Society list in 1969 where he spent five short seasons. He did not leave a strong presence behind him when he moved to Claregalway in 1974. One of his four registered sons, Silver Finn out of Finola of Leam,[92] joined the Killyreagh stud in the North of Ireland. Many of his one hundred registered daughters were true to type and went abroad. Perhaps the Society should have persevered with him for longer.

In 1970 Ardilaun was joined on the Society list by another of Carna Bobby's sons, Killyreagh Kim, and he remained there until the end of 1986. Killyreagh Kim was not the most attractive youngster but with old age he acquired a certain dignity and stateliness. His strong frame had prevailed against the rigours of his climate, and occasionally indifferent care. He displayed the best qualities of his breed in his twilight years, which were spent in Tommy O'Brien's good care. Many of his progeny, of both sexes, have proved valuable breeding stock. It is certain Killyreagh Kim will always hold a special place in the Societys stallion roll of honour.

Only recently did I discover Killyreagh Kim's roots. For some strange reason I did not look upon Killyreagh as a prefix and, if I had, I had no notion from whence it came. Thanks to Out of the Mist I met Alice Crichton, and learnt how she and her husband, Colonel Michael Crichton, bred Connemara ponies at their Killyreagh home near Enniskillen, Co Fermanagh, Northern Ireland. Two of their

[91] S.I.M. 159 O.O.M. (see Index) [92] O.O.M. 26

foundation mares came from the renowned Ballydonagh stud and were daughters of Cashel Kate,[93] no less. Ballydonagh Kate by MacDara[94] travelled North carrying a foal to Carna Bobby, a colt, who was named Killyreagh Kim. Michael and Alice were anxious to forge good relations between breeders in the North and those in Connemara, where they always experienced kind hospitality when they travelled to Clifden. As a gesture of thanks and goodwill, Killyreagh Kim was offered to the Society and as a two year old he was taken to Graham Tulloch for safe keeping. He remained at Shanbolard until he was registered and ready to take up stallion duties in Connemara. Twenty-five years later we are well aware what a valuable gift he proved to be.

Between 1971 - 1975 twenty-one of his sons were registered. Two of them remain in Connemara today, Loobeen Larry[95] at Cleggan Farm and Sarsfield[96] in Spiddal. Both are privately owned. Loobeen Larry is as wise and kind as his sire and does a useful job over the smaller mare. His hard colour is a distinct advantage and from dun mares he often throws attractive dun daughters, a favoured colour which is in danger of being lost in Connemara. Larry has a dun grandson, Coral Strand, with the McDonaghs in Carraroe.

Sarsfield spent five seasons on the Society list before moving to the Milford stud in Tuam. One of the fruits of this visit is John McLoughlin's lovely brood mare, Milford Wren. When I was travelling the inspection route in the spring of 1990 there were several progeny by Sarsfield in the Spiddal area and I discovered he had returned to Connemara. In 1992 I met him at his present home, Cregduff. He was in his twentieth year and more reminiscent of his sire than Larry. He no longer carried the dapples of earlier years and was 'spare' but bright of his eye and in regular use by local breeders. As I fondled his nose over the stable door I reflected on his notable namesake, one Patrick Sarsfield, the romantic hero of the Williamite Wars in the seventeenth century. That very morning a programme on RTE was reliving the battles of Sligo, the Boyne, Athlone, Aughrim and Limerick. I realised for the first time Sarsfield was not a place name but that of a noble person. There was something of nobility too, in the bearing of this old pony who does not expect too much of life but continues to serve his native breed faithfully.

Larry has a son standing beside him at Cleggan and Sarsfield has one in Co. Clare, Kilduffahoo. Now that Killyreagh Kim is no longer alive, a good son of Sarsfield or indeed of Windy Day would surely be welcome in Connemara. A quick leaf through the most recent Volume of the stud book indicates there are as many progeny registered by Sarsfield as any other between 1987 - 1989.

Carna Bobby has forty-eight registered sons, an unequalled record. The majority of them were sired after he left Connemara and have ensured his contri-

[93] S.I.M. 215 O.O.M. 206 + 213
[94] S.I.M. + O.O.M. (see Index)
[95] O.O.M. 276 + 277
[96] O.O.M. 271

bution to the breed has been well distributed throughout the world. Among the good sons who remained in Ireland were The Admiral, two Ballydonagh ponies Bobby and Rob, Tantallon Bobby who, three years running, stood in second place at Clifden show, and Ashfield Bobby Sparrow who is still alive. There are sons and grandsons of those mentioned and many more. It is unlikely Carna Bobby's place in the history of the breed will ever be forgotten.

I am always presented with a dilemma when I write of Bridge Boy[97] who, as a son of Tooreen Ross,[98] belongs to another branch of the fruitful Rebel line. He was my introduction to the Connemara pony and the sire of one of my best mares and yet I cannot be whole-hearted in my appreciation of him.

In the 1960s Bridge Boy was the proto-type Connemara stallion, a tough little character who looked able to climb any mountain and live happily on its summit. In reality, he was stabled and breathed fire and brimstone when he appeared in the Clifden ring, where he was supreme in the decade of the sixties. He was privately owned by the Geogeghans and never moved out of Oughterard but was included on the Society approved list from 1963 - 1975. He has more than two hundred registered daughters and twenty-four stallion sons. Neither the Society nor Paddy retained a son for use in Connemara. Bridge Boy was not a consistent sire but, in his defence, I believe he covered many indifferent mares. His progeny so often lacked quality but were fertile breeders and as hardy as they come. Many have served the breed well in other lands.

I scraped the barrel to find male descendants in Ireland today, and discovered a son of Marco Polo of Clonkeehan, who was by Bridge Boy, in Co. Antrim and a grandson of the same sire in Co. Clare. I so very nearly missed his greatest grandson, Ard Conneely, Clifden Champion in 1992. His breeder is the enthusiastic stalwart, Margo Dean of Co. Meath. As I sat at the Clifden ringside in 1991, Margo joined me on the concrete seat. True to fashion the ground was periodically swept by a sharp shower rolling in from the Atlantic. It prompted Margo to open her umbrella, nearly taking the right eye out of me as she did so. As the stallions paraded I remarked "that pony would be my choice", "Its mine" she exclaimed and up went the umbrella with excitement. I was in danger of losing both eyes. Ard Conneely was placed third on that occasion, standing between Tulira Finn MacCool and Smokey Duncan.[99] They dwarfed him in inches but not in build. The best compliment I can pay him is to say his grandsire has good cause to be proud of him.

Ard Conneely has May Boy[100] and Adventure[101] on the top line of his pedigree and Carna Bobby, Clonkeehan Auratum[102] and Dun Lorenzo[103] on the bottom. It would be difficult to find a sire carrying a better blend of the blood lines available.

[97] O.O.M. 187 + Index
[98] S.I.M. 114 + 160 O.O.M. (see Index)
[99] O.O.M. 278
[100] S.I.M. 192-194
[101] S.I.M. 142-144
[102] S.I.M. 195-198
[103] S.I.M. 169 + 170

While Tooreen Ross does not have a representative in Connemara today, the line is well secured in Co. Galway by the Tulira sires. They are a very different brand to the Bridge Boys with more size and scope. The Society were not lucky when they tried to introduce this line to Connemara in 1976.

As I ran my finger and my eye along the tendrils of the male Rebel network I discovered a second son of Inver Rebel, Greaney Rebel, who spent the last years of his life with Henry Whyte. Bred at Carraroe in 1963 he was registered by Martin McNamara in Co Clare, and won the Clifden stallion class for him in 1968. He was with Henry and on the Society list 1978 - 1982. Three sons and sixty-five daughters were registered but the male line was not sustained. I searched for a descendant of Strongbow, another interesting sire from the Rebel line, and found Moyglare Samson who stood in Co. Cork, where he has left two sons with the Templebready prefix.

This short survey shows how the Society depended on Rebel Wind and Killyreagh Kim and more recently Mervyn Kingsmill. The Rebel line has proved an enduring one and its importance is lost to no one.

CONNEMARA BOY DUN LORENZO LINE (2)	MACDARA	15 yr. old son of DUN LORENZO
	DUN AENGUS	6 yr. old son of DUN LORENZO
	ATLANTIC STORM	4 yr. old son of MACDARA
	BEN LETTERY	3 yr. old son of GAEL LINN

The Dun Lorenzo line had a very strong hand in 1964. MacDara, who was fifteen years old at the time, and had been on the list since 1952, had a son and a grandson on the list beside him. A second son of Dun Lorenzo, Dun Aengus,[104] was also there. They were strategically placed. MacDara was at Bunowen, Dun Aengus at Claddaghduff, Atlantic Storm[105] at Carraroe and Ben Lettery[106] at Ballinafad. MacDara's long reign provided thirty-three registered sons and two hundred and sixteen daughters. This was a good average for any of the long serving sires in Connemara. His line has been considered of prime importance by the Society, who has guarded and supported it more generously than others. Four of his sons have been on the approved list at different times, Gael Linn, Atlantic Storm, Ardnasillagh Magnus and Ormond Oliver.

Gael Linn, his first son to be registered, was privately owned by Tommy McDonagh and spent 1958 and 1959 on the Society list before leaving for the U.S.A. It is likely his presence would have passed unnoticed were it not for his son, Ben Lettery, who served the Society for twelve seasons before moving on to

[104] S.I.M. 80 + 172 O.O.M. (see Index)
[105] S.I.M. 173 O.O.M. (see Index)
[106] S.I.M. 173 O.O.M. (see Index)

Lambay Island. Golden Dan[107] is the only one of his eight registered sons to render a significant service, which he has done most gallantly in Germany. Ben's two hundred and five registered daughters were workman-like individuals whose hardiness was a valued quality. His progeny rarely appeared in the show-ring and neither did he. Ben Lettery was, I believe, a classic example of 'handsome is as handsome does'. He and others like him have provided essential backbone to the breed, without which it would be the poorer.

Atlantic Storm is recognised as MacDara's most prestigious son. When I was in France in 1990 I was asked by a French breeder why I had such a poor opinion of him. I am sorry that two lines in Shrouded in Mist[108] gave this impression. A closer inspection of his progeny confirms my suspicions he may not have entirely fulfilled the expectations of him. He was on the stallion list 1964 - 1977 and spent his last years in Carlow. He is represented today by a son, Beal Tra, who was in the Carraroe region for a spell before moving on to Co. Cork. A second son, Quiz Master, was used at Tulira briefly and has left a son, New Beginnings, who is with Willie Leahy at Loughrea. The last of Atlantic Storm's sons to be registered was Granard Storm. He spent one season in Connemara and is now with Clive Evans in Donegal. A son of his, Ashfield Festy,[109] was on the Society list 1989 - 1992 and he has left some attractive fillies in Connemara. I suggest that Atlantic Storm did not stamp his sons with authority and that he reserved his best qualities for his daughters, many of whom are excellent brood mares.

MacDara's third son, Ardnasillagh Magnus, is a mystery. He spent four seasons on the list and was with Paddy Geogeghan at Canrower who did not rate him very highly. He sired eleven registered daughters before leaving for France.

Ormond Oliver[110] was sired in the last and twenty-fifth year of MacDara's life and was bred in Co. Kilkenny. He was bought by the Society and was on their list for ten seasons 1977 - 1987. I believe he was not an over-popular sire while in Connemara. Maybe it was his golden dun colour which deterred breeders. He was produced in great fettle for Clifden in 1984 by Paddy Folan and stood second to Killyreagh Kim. His last two seasons in Connemara provided only six registered daughters. He too, went to Carlow on his retirement, and a small trickle of progeny continue to be registered. He has a son, Newbridge Oliver, with Mick Laffey in Clifden and another, Ashfield Mactalla, has recently travelled to England. A further Society owned son of MacDara, Maam Hill, was used in Connemara 1982 -1987. There appear to be very few of his progeny there today and he has moved to Co. Tipperary.

I hope I will not be considered a heretic when I say I find Dun Aengus a more profitable sire to follow. Both on pedigree and looks he was an unlikely

[107] O.O.M. (see Index) [108] S.I.M. 173 [109] O.O.M. 277 + 278 [110] O.O.M. 273

choice for the Society to make. However, he proved his mettle between 1962 - 1977, his greatest contribution being his sons, sixteen of whom were registered. I remember being advised he was the one sire to avoid when looking for a colt in 1973. Perhaps I was poorly advised. Two good sons went overseas, but Kimble,[111] the third to do so, left Connemara a rich legacy from which breeders benefit today.

Dun Aengus was named after a mountain on the Island of Arran and a boat of the same name which provided a passenger service between the mainland and the Island. He showed none of the strength of his namesakes in his person, but, apart from one year, when he was off-sick, he withstood the same harsh life as that of any other stallion who served the breed in Connemara. His son, Kimble, was privately owned by Willie Diamond and spent five seasons on the approved list. He was second to Bridge Boy at Clifden as a four year old and won his class in 1970 and 1971. He showed over-much quality for some, but his conformation, his head, his bone and his character were hard to fault. His two good sons, The Fugitive and Abbeyleix Owen, certainly show greater substance and size than their sire. Kimble's twenty-nine registered daughters do not surface very often.

Abbeyleix Owen,[112] like many another, did not come into his own until mid or later life. Today he is held in high regard in Connemara and it is regrettable that his good son Cuchulainn[113] died at an early age. Cuchulainn has left two useful sons, Coral Star at Derrygimla and Slisenoir at Errisbeg. Abbeyleix Owen has further sons, Aran Flight in Moycullen and Pilgrim in Loughrea. I wonder if the Society have it in mind to retain one for their own use? Owen is also represented by his son Callowfeenish Mairtin in Co. Wexford and his half-brother, The Fugitive,[114] is in Co. Meath. They provide a good back up service to this tried and tested blood line.

Dun Aengus leaves one further son who should receive a mention, Calla Boy*. He has always been privately owned and lives at Ganty where it is likely he has not received many mares. It would be easy to overlook him, but at eighteen years of age might he not still have a role to play somewhere?

I hope my review of the Dun Lorenzo line reflects my esteem for it. The future of the MacDara branch of the line appears to be in some jeopardy and it must be hoped a really good male will present himself. Ormond Oliver is still alive and Newbridge Oliver at only five years of age has plenty of time yet to prove himself. As I write the Society do not own a sire from the MacDara line. New Beginnings, a great-grandson, covers a good number of mares each season at his home in Loughrea. A bright new star will surely surface somewhere to supply the future needs of the breed.

[111] O.O.M. (see Index)
[112] O.O.M. 275 + 276
[113] O.O.M. 275
[114] O.O.M. (see Index)
* See ch. seven

CARNA DUN	CARNA DUN	16 yr. old son of LITTLE HEAVEN
LITTLE HEAVEN	DOON PADDY	11 yr. old son of CARNA DUN
(T.B. LINE) (3)	MERVYN STORM	4 yr. old son of CAMLIN CICADA

Readers will not be surprised I feel more at home as I move onto the Carna Dun line. As a breeder, it is unlikely I would have used any of the sires from this line, but it is equally certain, the breed could not have done without them. By and large they were on the 'horsey' side, often with plain heads. Nonetheless, Carna Dun's forty-eight registered sons and two hundred and fifteen daughters provided a valuable outcross to the Rebel and Dun Lorenzo native lines.

How often I find it is the first son who is given the opportunity to follow his sire on the stallion list. This was the case with Doon Paddy[116] who, on paper, appeared an admirable choice. His dam, Doon Lily, came from reliable foundation stock. Even with hindsight it would be hard to say the Society could have made a wiser choice from sons sired in his later years. Doon Paddy had a long life and while he was not the most handsome of ponies, he provided two good sons from the eight who were registered. They are Glyntown Paddy and Silver Rocket.

Glyntown Paddy was born in Co. Cork in 1963. His dam, Truska Colleen undoubtedly travelled from Connemara carrying him. She came from one of the very best female families and proved this by nicking so well with Doon Paddy. Glyntown Paddy, in his turn, has provided two registered sons in Ardravinna Andy who is with Henry Whyte in Salthill and Ardravinna Finnian who has remained with his breeder in Cork. Andy is playing a very useful role as a sire in Connemara in the 1990s and already has a registered son, Rynn Richard.

Silver Rocket was bred at Callowfeenish, but has spent his life in Loughrea where he has been regularly favoured by breeders. His dam, Slieve Winter, came from an old reliable family, but as her sire was Creganna Winter,[117] Rocket shows T.B. blood on the bottom as well as the top line of his pedigree. A son, Seafield Silver Rogue, was bought by the Society for their use in 1987 and moved to Co. Donegal at the end of 1992. I have not been an admirer of this pony but I notice a son of his, Little Rascal, won his class at Clifden in 1992. Only time will tell the value of this line. Silver Rocket, (who has a grandson, Gorteen Boy, in Spiddal) and Glyntown Paddy have, between them, ensured that Doon Paddy's presence remains in Connemara. Two further sons were not so successful. Lough Eskey was on the Society list for a single year in 1972 and Abbeyleix Cypress was used in Co. Wicklow briefly before being gelded.

A second son of Carna Dun, Camlin Cicada,[118] provided a son for the Society in Mervyn Storm,[119] who was on their list for five seasons, the fruits of which were

[116] O.O.M. (see Index) [117] S.I.M. 186 [118] O.O.M. (see Index) [119] S.I.M. 191

five registered sons and ninety-five fillies. I have never been able to discover why he was discarded so soon, spending the remainder of his life as a teaser in Co. Meath. Two of his privately owned sons, Clare Charger and Rosaveel Storm, were used in the Carraroe region for a while. I have often admired good daughters by Mervyn Storm and would have been happy to own one.

Another Society owned son of Carna Dun, Toombeola,[120] was on their list in 1968 for two seasons. He provided forty registered daughters in that time. Beola was a giant or chieftain whose name marked his tomb, Toombeola, and likewise the mountain peaks, Benna Beola (the Twelve Bens). It would have been fitting if the sire carrying this strong, historic name had made his mark in the pony world.

Camlin Cicada was bred by Garnet Irwin, but spent all his working life at the Clonkeehan stud. He was an upstanding bay pony and two of his sons, Checkpoint Charlie and Clonkeehan Nimbus, made a worthwhile contribution. Westside Fred, a son of Charlie, lives beyond Galway at Turloughmore but this does not deter breeders who live in Connemara from making the long journey to visit him with their mares. I saw Clonkeehan Nimbus in 1977 and found him to be a strong, small horse who was well used for cross breeding and who sired animals with outstanding jumping ability. Nimbus supplied the Society with a son, Skryne Bright Cloud, who was used in Connemara for five seasons 1977 - 1981. His contribution was a modest one, with two registered sons and forty daughters. A son, Robin Hood, (Village Grey) stands in Corundulla. Jimmy's Fireball, also by Nimbus, spent a number of years in Moycullen. I can hardly pass on from Carna Dun without special mention of two further sons, Little Joe and Kilgreaney Lad. I will catch up with Little Joe later.

Kilgreaney Lad was bred at Lissoughter near Recess in 1965. He was a great-grandson of Retreat[121] and I regret his grand-dam Ruby, by May Boy, was mistakenly shown as having no progeny in Shrouded in Mist. Lad was registered by John Finn of Athenry and later spent some time with Eddie Madden, also of Athenry, before moving on to Murty McGrath of Claregalway. It was Murty who bred three useful sons by Kilgreaney Lad, Cregmore Kiaran, Smokey Shane and Smokey Fireman.

Kilgreaney Lad moved onto John Daly in his old age. By this time Murty had two of his daughters and needed an out-cross sire to use on them. It is interesting to note how this sire was valued and passed from one private owner to another. He left seventy-two daughters and eight sons. He was grey but he carried the valuable 'hard' gene and many of his progeny came bay. He has a son, Dimples, in Co. Monaghan and another, Island Lad, in Co. Mayo.

[120] O.O.M. 68-77 + 271

[121] S.I.M. 216-217

Carna Dun's fine reputation rests primarily on the performance of his daughters as brood mares but this short review proves he has been a valuable 'all purpose' sire. His sons may lack the quality and looks of his daughters but their record, as sires, bears a close look. Carna Dun was a courageous pony, who was, after all, a half-bred with warm blood in his veins. And yet he lived the same hard life of all the native sires, often hobbled, with little shelter, travelling the roads beside a bicycle and covering as many mares in a day as he was asked.

CLONKEEHAN	CLONKEEHAN AURATUM ..... 10 yr. old son of NASEEL
AURATUM	CLONJOY .... 3 yr old son of AURATUM
NASEEL	AURA DUN .... 5 yr. old son of AURATUM
(ARAB LINE) (4)	KNOCKRANNY RUBY .... 3 yr. old son of AURATUM

Clonkeehan Auratum[122] appeared well represented in 1964 with three young sons on the list. It proved to be a false dawn ... Knockranny Ruby belonged to Peter Kyne and stood beside his sire Auratum, at Knockranny, Moycullen, for the five years he spent on the Society list. He provided two registered sons and forty-eight registered daughters. One of the sons, Aura Dun, went to the U.S.A. Before leaving Connemara he spent 1963 in Rosmuck and 1962 and 64 at Oranmore. During my research I have come across some good daughters of his and wonder, with hindsight, if it was a mistake to part with him so soon.

It was left to Auratum's son, Clonjoy,[123] to keep the line alive in Connemara, which he did in the best possible way. He was on the society list from 1961 right through to 1972, spending his semi-retirement years with Mr. Quinn at Ardrachan and leaving twenty-three registered sons and two hundred and two daughters. It is him I have to thank for Island Duke. Two of Clonjoy's sons were on the list, each for two years; Errigal Prince for 1965 and 1966 and Rosmuck Master for 1974 and 1975. Errigal Prince provided seven registered sons and thirty-five daughters. One son, Abbeyleix Ri, was bred by Susan de Vesci and spent his life in Co. Laois. He it is who supplied a tiny trickle of Clonjoy's blood which today flows through the veins of the Society stallion Smokey Duncan.[124] Ri was the sire of Bobby Brown (now in the U.S.A.) Bobby Brown was the sire of Westside Frank[125] (now in Germany) and Frank was the sire of Smokey Duncan. How very nearly this tiny trickle escaped Connemara.

I believe Errigal Prince may have been under-used and under-rated. After he left Connemara he made a name for himself as a successful show jumper while with the Connolly family in Ballyglunin, South Connemara. Many of the progeny from the Errigal Prince line show this same ability. I doubt if Rosmuck Master had anything special to offer. I would like to have seen a further son of Clonjoy's at

[122] S.I.M. 197-198 + O.O.M. Index [123] S.I.M. 198 + O.O.M. (see Index) [124] O.O.M. 277-278 [125] O.O.M. 174 + 175

work in Connemara. Maybe he is not quite as good a sire as I like to think, but to this day whenever a filly in Connemara takes my eye, Clonjoy is invariably somewhere in the pedigree. Proof surely, that he leaves his own mark, especially in the head. I have been unable to find evidence of registered progeny by Roobeen Boy, a son of Clonjoy foaled in 1966 and still alive in 1993. He has spent his entire life in Co. Kerry.

Towards the end of his life, Auratum sired Ashfield Alex[126] out of Lambay Starry Eyes, a Carna Dun mare. He was a small bay pony lacking in substance who spent seven seasons on the Society list without, it would appear, making a substantial contribution to the breed.

Clonkeehan Auratum did not prove as successful an out-cross as Carna Dun. However, his presence broadened the genetic base at a time when it was urgently needed, and he left sufficient good progeny among the mediocre ones for which many of us will always be grateful.

Last, but by no means least, comes the fifth male line spear-headed by Mountain Lad,[127] and mentioned earlier in this review. There was no male representative at work in Connemara in 1964, but it was re-introduced in 1969 after twelve years absence. Tully Lad, a son of Mountain Lad, left the Society list in 1956 and he was not replaced. The line was conspicuous by its absence from the Society list for twelve years. Luckily John Brennan retained a son of Tully Lad, Thunder, and it was John who offered his son Thunderbolt[128] to the Society. He was registered in their name but did not appear on the list until he was six years old in 1969. Where did he spend the first six years of his life I wonder? He was never truly accepted in Connemara although the value of his blood line must have been recognised by some. He spent 1969 - 1972 on the list before going to Limerick as a teaser. He returned in 1980 for four seasons before travelling to England in 1984. The Society have used three of his sons; Thundercloud for four seasons, 1973 - 1976; Thunderstorm for one in 1986; Canal Cormac for two, 1987 - 1988. There is no son of Thunderbolt's in Connemara today but Thunderstorm is alive and well in Co. Meath and Canal Cormac has a son, Thunderball, in Co. Mayo. The Society's endeavour to keep this line on track in Connemara has been a diffident one and not entirely successful. On the credit side, a current Society stallion, Moy Hazy Cove, is a grandson of that great little jumping sire from Co. Longford, Tully Grey,[129] who was also a son of Tully Lad. Moy Hazy Cove has both bone and substance and, hopefully, will breed well over the quality Kingsmill mares.

I have to be careful not to overplay Thunderbolt in my regard for him and the line he represents. I confess to disappointment at the apparent failure of his

[126] O.O.M. 254 + 274 [127] S.I.M. 177 [128] S.I.M. 180 + 205 O.O.M. (see Index) [129] S.I.M. 179 + 180 O.O.M. 23 + 189 + 239

sons in Connemara and I ask myself were they not good enough or were they not given long enough to prove themselves? I accept that Thunderbolt is a sire who needs to be used selectively and this would not have been practicable in Connemara. I have every confidence this virile line will survive in its homeland given only half a fair chance.

I have enjoyed delving into old stud books, old records and old memories. I hope this review will appear as impartial as I have endeavoured to render it and that there are no glaring omissions.

* * * * * *

I had reached desperation point as I began work on Chapter Eleven, Selected Mares, for Shrouded in Mist. This was way back in December 1981 when my three month sabbatical was drawing to a close. The pressure to complete my task and the weight of my over-taxed brain were almost too much to bear. Quite suddenly the load lifted as I became immersed in the treasure trove of mares assembled before me. They restored confidence in myself and in the Connemara pony. There was no time to plunge their depths, nor even now is that possible. On many occasions in the last ten years I have returned to pages 203 - 230 to record an exciting new discovery in their midst. The mare family charts are criss-crossed with many coloured pencilled additions. Now is the time to share them.

In 1985 I received a courteous letter from Bartley Walsh of Kylesalia. He was anxious to offer corrections to my piece on the mare Cuach na Coille.[130] She was, he told me, bred by the Connollys of Kylesalia and bought from them by his father, Martin Walsh. She was believed to be by Charlie[131] and was sold for £17.00 not £35.00 as suggested in Shrouded in Mist. Cuach na Coille, correctly translated, is 'Cuckoo in the Woods' and she was probably so named because she came from the heavily wooded area overlooking the cove which separates Kylesalia from Loughconeera. Bartley went on to tell me the Walshs of Kylesalia kept a stallion from 1903 - twenty years before the birth of the Society - until the early 1960s, firstly by his grand-father Paij Walsh and later by John Walsh. Barley's letter concluded "Thank you for a very enjoyable book. God Bless you in your work." This letter is among my treasured archives. In 1985 I visited Kylesalia to meet Bartley and to thank him in person.

Meantime I have discovered I owe Cuach na Coille an apology for dismissing her as a brood mare of little worth. In 1944 she bred a filly foal to Gil,[132] Cuach Og who in 1951 produced a filly by Calla Rebel,[133] Jump the Wall. She was sold to the Mulkerrins of Callowfeenish for whom she bred five good fillies all by different

[130] S.I.M. 212-213 O.O.M. 8 [131] S.I.M. 141-142 [132] S.I.M. 155 + 156 + O.O.M. (see Index) [133] S.I.M. 154 + 155

sires. The second of them, Homeward Bound by Rebel Wind remained in Callowfeenish and bred three fillies, Home Lover by Clonjoy, Party Gay Girl by Ben Lettery, both of whom went to Maria Levinge, and Knockabout Lady by Dun Aengus. I found Knockabout Lady with Bartley Walsh at Kylesalia where she has been since registration at two years. Thus the two families Walsh (human) and Cuach na Coille (equine) are reunited some three and four generations later. In 1993 Bartley registered a two year old filly by Mervyn Kingsmill, a granddaughter of Cuach Og (the Little Cuckoo) and he named her Cuach Glas na Coille (Grey Cuckoo of the Woods). She will remain at Kylesalia beside her dam Knockabout Lady.

Morgan Mulkerrins mare, Home Lover II by Thunderbolt, ensures this line continues to breed at Callowfeenish. Her filly, Callowfeenish Star by Ashfield Festy was forward for registration in 1993.

Jump the Wall ended her days in Carlow with Jimmy Jones where she bred a filly to Carna Bobby, Ashfield Judy, and another to Thunderbolt, Ashfield Rolling Rock. Judy is the dam of Ormond Oliver and Rolling Rock went to France.

Cuach Og's second filly by Dun Orphan was bought by Tom Hynes at eighteen months for £18.00. Tom walked the filly the many miles from Kylesalia to Canal Stage. Two years later he sold her to Jack Bolger for £85.00., which, in 1958, was a tidy profit. Jack registered her Moonlight Rambler and sold her to Stephen Mullins of Ballyconneely where she bred ten foals. A colt by Carna Dun, Ballintemple Dun Rambler, went to Northern Ireland and is the sire of the successful stallion in France, Knockadreen Holiday Cliff. In 1991 Tom Hynes son, Padraig, judged at Tours and presented the Challenge Trophy to Francois Birkigt and Cliff. Rather a nice conclusion to this pony story.

What a discourtesy to Cuach na Coille to suggest she was not a good brood mare! It only required Cuach Og to prove me wrong. How happy I was to find this pony family in good heart, where it belongs, in Kylesalia and Callowfeenish.

It would have required more than one good polish of my crystal ball to foresee the rich fruit the branches of the Slieve na mBan[134] family would provide. it has flowered in variegated blossoms for the benefit of breeders in England, France and Sweden but sadly is no longer breeding for the Folans at Callowfeenish.

Sliabh na mBan 2nd by Calla Rebel was foaled in 1952 and I notice her son Calla Boy was foaled in her twenty-sixth year. I wonder how her dark brown daughter by Rebel Wind came by the name of Mountain Silverwing? It was Silverwing's daughter Slieve na mBan Peggy by Killyreagh Kim who produced the 1992 Clifden winner, Sliabh na mBan Cailin by Abbeyleix Owen. It is good to know Padraig Hynes has Cailin at Canal Stage and that, since the Folans dispersed their mares, this valuable line continues to breed in different areas of Connemara.

[134] S.I.M. 210 + 211 O.O.M. 229 + 277

A grand-daughter, Miss Charger, has proved a good brood mare in the Clifden region and another happy story is the rescue of Callowfeenish Beauty, a golden dun daughter of Clonjoy and Sliabh na mBan 2nd. In 1989 she was taken to Maam Cross Fair with little flesh on her bones and little likelihood she was carrying a foal to Murphy Rebel with whom she had been running all summer, on a small off-shore island. Jimmy Canavan took a chance, and bought her for his son John. She produced a healthy foal next spring and two more since. Beauty's first registered daughter, foaled in 1971 by Ben Lettery, is the dam of Callowfeenish Thunder[135] (U.S.A.) and Newbridge Oliver stationed in Clifden in the 1990s.

Callowfeenish Pride hails from another branch of this remarkable family and her daughter, Wireless Wave (no relation to Wireless Wave C.P.S. 477), by Clonjoy has produced some good ones such as the sire Callowfeenish Mairtin and the mare Cailin Ciuin, Frank Joyce's good Clifden winner. Wireless Wave is no longer at Alnabrone but she has left a daughter there by Tantallon Bobby.

It should be said, that, once again the 'caption' Callowfeenish has been used at will by many breeders in the Callowfeenish region. The redoubtable Callowfeenish Dolly II[136] is one who has no connection with any of those already mentioned. The Devanes, who bred from a Carna Dun mare, Vera, also used Callowfeenish when naming their ponies. Callowfeenish, which lies midway between Carna and Kilkerrin covers a wide area of scattered homesteads and small farms. The Mulkerrins - and there are many branches of this family - the Folans, the Walshs and the Devanes have long been loyal pony breeders in this region. They are diminishing in number, as are the ponies, and it would be a sad day for the breed if they were to disappear from this little enclave altogether.

As I prepared to tear myself away from the Sliabh na mBan family chart, my eye fell on the simple name of Katie. Immediately a memory bell tinkled. Could this be the Katie I met in Australia, dam of Yarraman Park Henry,[137] the sire who took my fancy? Yes, it was and she, Katie, is a half sister to the stallion in Loughrea, Silver Rocket. It goes to show how the fruitful branches of this good family have spread themselves far and wide.

Retreat's[138] chart is greatly enriched by the discovery of her grand-daughter Lissoughter Dolly, daughter of Ruby and Carna Bobby. Dolly is not only the dam of Kilgreaney Lad but of Lissoughter Laura by Clonjoy, foaled in 1973. From Laura's extended pedigree I find she breeds back to her great grand-dam Retreat through both her sire and her dam. Sceptics may raise an eyebrow and ask is that not 'in-breeding' upon which you frown? My answer would be "No". I regard this as an acceptable route to take, breeding back to a good mare supported by good outcross sires on either side. Laura has Silver Pearl[139] and Clonkeehan Auratum on

[135] O.O.M. 109 [136] S.I.M. 230 O.O.M. 263 [137] O.O.M. 146 [138] S.I.M. 216 + 217 [139] S.I.M. 175 + 176

the top line and May Boy and Carna Bobby on the bottom line. Plenty of variety there I would say.

As a teenager in the hard years of the forties and fifties, Christy Sullivan rode Retreat bare-back many miles a week. He remembers well her qualities, both as family work-mate and regular breeder. I hope Lissoughter Laura may still be with Christy. If not, I'm sure her daughter Lissoughter Cora by Ormond Oliver will be roaming the 'Fairy Mound,' behind the Sullivan homestead, with a foal at foot. (It is thought Lissoughter indicates the presence in times past of a Ring Fort or Fairy Mound in this area).

I have the photograph of Cashel Kate[140] in front of me. The longer I look at her the more certain I am she was by Carna Dun and not Tully Lad, as the stud book tells us. She was foaled in 1956 at Kilkerrin and in 1955 Carna Dun was standing up the road at Kylesalia, while Tully Lad was stationed at Roundstone, a good cross country step away. Kate's breeding record as a brood mare is typical of a good Carna Dun mare. From her eight registered progeny she had four sons and four daughters, four of whom were by Carna Bobby. Her sons, The Admiral and Ballydonagh Rob in Ireland and Sticky, for the greater part of his life in Holland, have given valuable service to the breed. Their full sister Ballydonagh Deirdre has carved herself a unique place in Connemara history. She left the Ballydonagh stud in Co. Meath at the age of two, carrying a foal to Mervyn Invicta, and travelled to Biddy Scott's Mullaghmore stud in the North of Ireland. She later moved nearby, to the Crighton's Killyreagh stud. In 1971 she won the five to six year old mare class for them at Clifden, and stood reserve to Atlantic Surf for the Archbishop's Cup. She visited The Fugitive while in the South and produced a filly to him the following spring. My detective work suggests she was bought by Frances Lee Norman when the Killyreagh stud was dispersed. In 1979 she produced a filly foal by Clonkeehan Yeoman, in Holland for her new Dutch owner, Mr. Van der Meijden.

In 1992 Anne Hemphill and I were invited to Judge at the Dutch Society's 25th Anniversary show at Heteren. We chose twenty-eight year old Ballydonagh Deirdre, looking as fresh as a daisy, as Supreme Champion of the show. In my view she represented a perfect miniature middleweight hunter but required a splash of 'pony paint' to be a perfect representative of her breed. None the less, there was not another who could hold a candle to her and it was a proud day for all her connections. She had three of her progeny forward for the family group: her son Marvel, by Carna Gold,[141] who was Champion stallion, and her two daughters Diana by Atlantic Curragh[142] and Suzan by Clonkeehan Yeoman. They all carried Mr. Van de Meijden's Van Graaf Janshof suffix. Alice Crighton tells me her half

[140] S.I.M. 214 + 215 [141] O.O.M. 208 + 209 [142] O.O.M. (see Index)

sister, Ballydonagh Kate by MacDara, was not in the same class and bred only the one good one, Killyreagh Kim. There are daughters and granddaughters of Cashel Kate breeding in Ireland today. I very much hope Deirdre's two Irish bred daughters Mullaghmore Siskin and Killyreagh Jane, may not be too far away from their birth place.

Little Nell[143] had thirteen daughters registered between 1949 - 1968. This was a rare contribution by any standards, but for a mare living on the Truska beach it was an extraordinary one. Her owner, Val Gorham, must be given some credit for this. I find he registered a grand-daughter, out of Fairy Legend, the eleventh of the thirteen, as recently as 1983. She was Pigeon, by Ashfield Alex out of Pigeon on the Strand. In 1992 I met ten year old Mary Ward at the Streamstown bridge inspection point. She had taken a day off school to present her filly out of Pigeon by The Kidd to the inspectors.

Lehid Rose, the first of Little Nell's thirteen fillies, was registered by Martin Conneely in 1951 and one of her family has been breeding at Lehid ever since. Rose lived into her late thirties and died in the field in which she reared so many of her foals. The Lehid ponies live alongside the Village ones, and may not enjoy the same world wide attention as they do. They are however a hardy, reliable breed and today it is twenty year old Lehid Lovely Princess by Killyreagh Kim who maintains the family tradition. In 1978 I brought Lehid Linda to England for Freddie and Rachel Grant as a wife for Chiltern Curlew.[144] She too was a daughter of Killyreagh Kim and had been a raffle prize pony rescued by Graham Tulloch. As I scan Little Nell's abundant family chart I realise all too many have gone overseas, but I rest content that Paddy Conneely and Mary Ward will ensure the family continue to thrive, not too far from Little Nell's breeding grounds.

I turned pages 203 - 230 of Shrouded in Mist back and forth. We have already seen the Winnie and Windy families are in safe hands. Carna Dolly may have only one grand-daughter breeding in Co. Wexford, and Lougheera none at all in Ireland. I stress, may, because it will be seen I am not infallible and ponies are often re-discovered after having been lost to the breed for some years. Equally secure is the Winter Roche family with Josie Conroy, Jane Dalrymple and others. I would be more fearful for Winter Heather's family which I saw dispersed at Maam Fair in 1978. Silva will surely have descendants in Co. Cork but have all Irene Grey's good Fort daughters gone overseas? I wonder if Canrower Lass has a grand-daughter round about the Geoghegan homestead? I think it is unlikely, but we know Arctic Moon's line continues in England. This leaves only Cregg Lassie[145] and what more is there to say about her? I have to register a hint of sadness that she does not have a good daughter breeding in Connemara today, though I am

[143] S.I.M. 224-225 [144] O.O.M. 53-54 [145] S.I.M. 228-229 O.O.M. (see Index)

aware they can be found in Co. Cork and overseas. But what of her sons, Little Joe and Cregg Hero, who are worthy of more than a passing glance?

I had long harboured a desire to meet Little Joe[146] and the Stagg family. Tales of his prowess on the race tracks and his close resemblance to his sire and dam, both firm favourites of mine, fostered this desire. Somehow the drive to Mayo seemed longer than it really is. I had never been able to justify a day away from Connemara when my time there was always so short. In 1990 Little Joe, in his twenty-seventh year, was invited to parade at Clifden Show. What a welcome he received and how well he belied his years. I promised myself that, next year, I would make the trip to Hollymount to meet him in person. We set out on the road to Mayo as if on a pilgrimage to meet a legend in his life-time, a legend indeed but with a very simple story to tell. We received a royal welcome and a feast fit for a king had been prepared for us. We ate as we relived Little Joe's life.

In 1963 James Walsh took Cregg Lassie and her colt foal by Carna Dun to the Oughterard Fair to sell the colt. Sean Stagg and his wife Veronica were on their honeymoon and chose a visit to the fair as a day out. I very much doubt they went with the intention of spending thirty pounds on a colt foal. Little Joe walked home to Cregg with Jim and Lassie that night and Sean arranged to pick him up the following day. He remembers arriving at Cregg in his 1956 Ford Anglia car. It had no proper hitch but somehow he had attached a very small borrowed trailer. It did not come up to Jim Walsh's standards for the transport of a foal and an agreement was made to borrow something more suitable from Paddy Geogeghan. Sean remembers it cost him a second round of 'luck money' but he was grateful for the neighbourly offer.

While Little Joe grew up in the care of his father, Sean and his wife went to England where Sean worked on the buses in Coventry. When they returned to Hollymount, Little Joe soon became an integral part of Stagg family life. He provided their sport and their pleasure and he helped to supplement their meagre income. He was a regular winner on the race tracks, be it Clifden or Ballyconneely. He made Grade A as a show-jumper and never returned from his local show at Hollymount without a rosette. On a Bank Holiday weekend it was not unusual for him to race on Saturday and Sunday, and then turn out in the jumping ring at Ballinrobe show on the Monday. I hasten to add he was not expected to do all this on thin air. Bone meal and cod liver oil were mixed in his shaved oats. On special occasions he was treated to Cannon Ball's old potion, a pint and a half of porter with six eggs!

His success as a race pony and show-jumper led to a steady demand for his services as a sire. Sean remembers the day when his weekly wage packet was £12.00, and yet he could earn £35.00 on a Saturday covering mares with Little Joe.

146 S.I.M. 190 + 229

On week days when he returned from work he found Little Joe prepared for an evenings travel. His father had a route mapped out from requests received during the day. His visits often extended way beyond Mayo to Sligo, Tuam, Clare and Galway.

Little Joe was more than friend to the family. He was their pride and joy. On one occasion they travelled him to Dublin for the show. There being no class for Connemara stallions he had to compete against all comers, in a class for stallions not exceeding 15 hands. He was placed fifth, with thoroughbred sires bred by the Queen and the Aga Khan at the top of the line. I would love to have been a spectator that day.

Our journey to Mayo to meet Little Joe and to share the Stagg's happy memories was well rewarded. We found him living contentedly in a meadow, once part of the Ruttledge estate, with a 'drying shed'* as his shelter. He still possessed the proud, full eye of his sire Carna Dun and, undoubtedly, inherited the gift of devoted service from his dam Cregg Lassie. It is a sad fact that the number of registered progeny by Little Joe in no way reflects the many mares he covered in his busy life. The greater part of them were thoroughbreds or pony mares of no known breeding. A good registered son with his courage and natural gifts would have been a bonus for the stud book.

A young Swedish enthusiast Jenny Hagenblad wrote to say how much she had learned from reading my two books and how they helped her to understand the Connemara breed better. She enclosed a photograph of a pony she had been riding named "Gregory". He was by Cregg Hero[147] and carried a good look in his person of his grand-dam Cregg Lassie. Cregg Hero's photo shows him to be every inch a pony and I could not help thinking it a pity he had not remained in Connemara to represent his father, Bridge Boy. His stud life in Sweden was all too short. He received a Grade B classification in 1969 and died in 1972 when only nine years of age. Jenny's long interesting letter concluded "Since I began learning more about the Connemaras a wish has been growing inside me and that is to once go to Clifden and watch the Connemara show. I hope that wish will come true one day." Jenny was shy about her English but she had no cause to be. I too wish we may meet at the Clifden ringside one day.

There are many unsung heroines in the C.P.B. stud book. A glance at a few who have crossed my path at different times provides a further insight into the origins of the pony we enjoy today.

As I wrote Out of the Mist Leam Lassie's[148] name often popped out of the pages. It was her remarkable Australian daughter, Ardan,[149] who first drew her to my attention. Between 1960 and 1974 eleven of Lassie's fillies were registered. You

* For drying timber

[147] O.O.M. 218 + 231

[148] O.O.M. (See Index)

[149] O.O.M. (See Index)

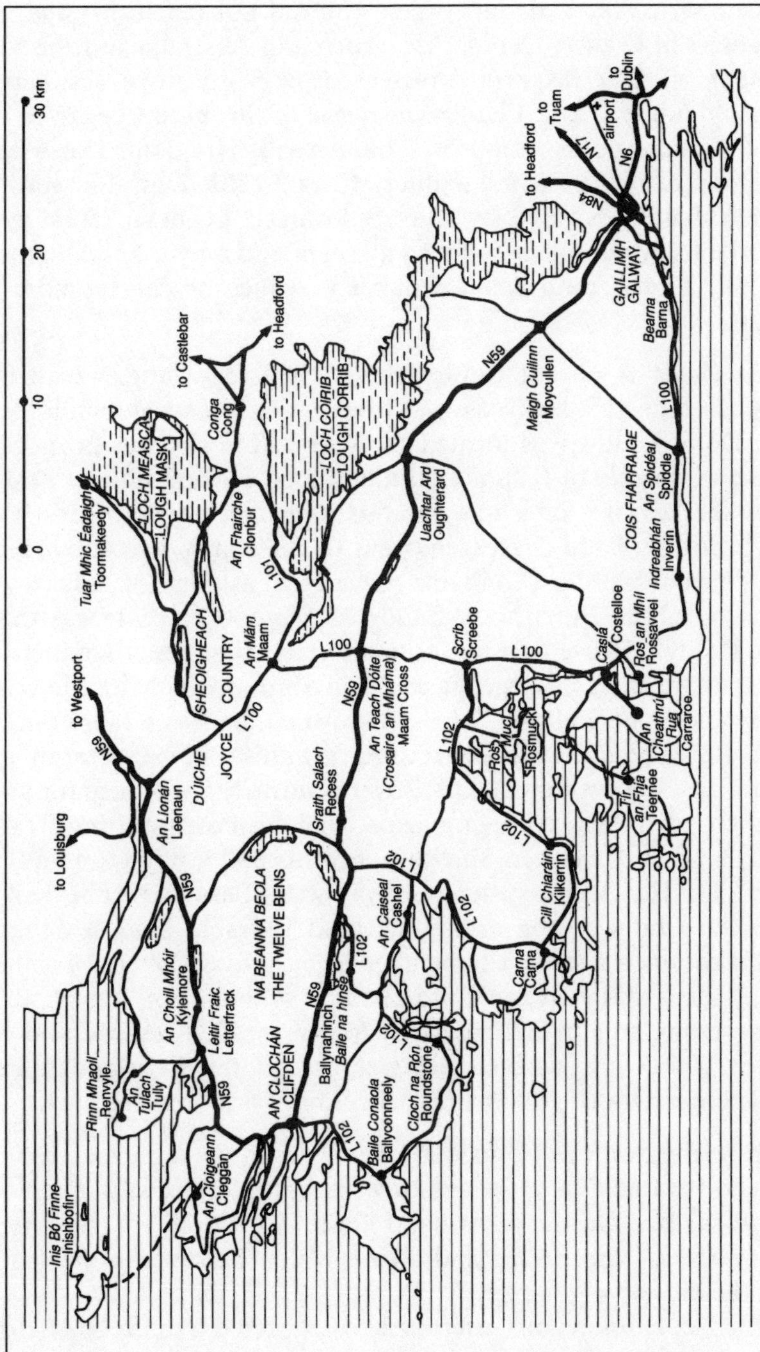

IAR-CHONNACHT / WEST GALWAY A BASIC BILINGUAL ROUTE MAP

(*Reproduced by the kind permission of Tim Robinson from his FOLDING LANDSCAPES studio in Roundstone*)

may depend there were colt foals inbetween who did not reach the stud book. I found her daughters in France, Denmark, Sweden and Australia and this may not be the full extent of their dispersal overseas. It was my hope she had left a daughter at her old home Leam. I had no evidence of this until I began to unravel the pedigree of Thunderbolt's grandson Thunderball. His dam, Dawn Mist by Ormond Oliver was, I discovered, a daughter of Leam Little Lady. She was the last of Leam Lassie's daughters, bred by John Sullivan at Leam in 1973, by Rebel Wind. Leam Little Lady continues to breed at Leam and one of her daughters was registered in 1993. Thunderball has a pedigree of which he can be more than a little proud.

Woodlands Rover is one of those strange 1960 registrations with her sire named but not her dam. It is likely she was bred in the Bunowen region for this is where her sire, Doon Paddy, was located the year before she was born. She was registered by Gerald Larkin of Killimore, Ballinsaloe. Her first filly by Bridge Boy was retained by the Larkins for whom she was a regular breeder. Rover's second filly, Westport Sandy, by Paddy's Heaven, went to Co. Mayo. I recently received a letter from a French breeder, Guillame Levesque, asking for advice on the prospects of her daughter, Burrishoole Sandy, as a brood mare. It was this letter which brought Paddy's Heaven to my notice for the first time. I regret he was omitted from the mare chart of Irene, page 205 in Shrouded in Mist. He was a son of Carna Bobby and Irene 2nd and, I have discovered, he was a more than useful sire. Woodlands Rover bred the Larkins a third filly, this time by Auratum and she went to Finland. In 1969 or thereabouts Rover returned to Connemara and bred three fillies in Letterdyfe before moving again, to John Folan on the Errislannan Peninsular. This is when I met her. She knew how to pull a hay-wain and how to breed a good foal. Her last registered filly, Sweet Lass by Abbeyleix Owen belonged to Eileen Conneely for whom she bred Bramble Lass, a daughter of Murphy Rebel mentioned earlier. I remember being offered her full brother as a yearling colt and shall always regret I did not take a chance on him. I wonder if Crackerjack, a gelding son by Thunderbolt foaled in 1984, (Woodland Rover's twenty-third year), was her last? Crackerjack joined the Errislannan trekking team. I'm sure there is a lot of good work in a gelding with breeding such as his.

Confusion reigned on the Connemara roads when the signposts reverted to the Irish. Costello became Casla, the twisting creek, and Screebe was replaced by Scrib, the end of the bay. On the other hand Doire An Fheich is more commonly Derrynea, the wood of the raven, and Leitir Mucu becomes Lettermacoo or Lettermuckoo, the hillside of the pigs.

In 1950 Michael Hopkins of Lettermuckoo registered a mare Grey Hop. Her name conjures the imagination of a small grey pony hopping over one stone wall

Ireland

1

2

1. Bobby Bolger and Coral Prince, 1993
2. Heather Wright and Ard Conneely, 1992
3. Sean Feeney and Lettermacoo Sweetie, 1987
4. Sian Carson and Grange Homing Pigeon, 1992
5. Abbeyleix Owen, 1991
6. Killreagh Kim 1992

3

4

5

6

N.B. Notes on the ponies featured can be found on page 146

Europe

7 **GERMANY**

10 **HOLLAND**

11 **NORWAY**

8 **GERMANY**

7. Bengt
8. Golden Dan
9. Bell Cazo
10. Marvel Van Graff Janshof
11. Palands Bajas
12. Knockadreen Holiday Cliff
13. Kasta Scaramouche
14. Frederiksminde Hazy Marvel

12 **FRANCE**

13 **FINLAND**

9 **SWEDEN**

14 **DENMARK**

England and the U.S.A.

15

In the U.S.A. 15. Balius Kerry Blue
16. Hideaways Erin Go Bragh

16

20

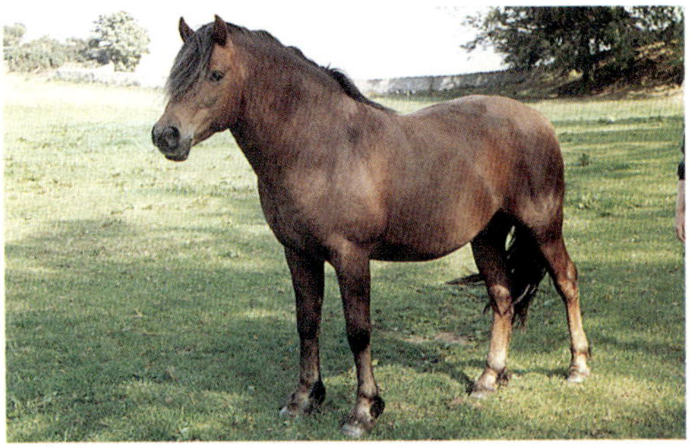

17

In England 17. Cocum Camelot
18. Blagdon Haze
19. Lewcombe Rondeau
20. Shipton Sea Imp

18

19

The Antipodes and Africa

21

AFRICA

21. Gilbert at Thaba Tseka, Lesotho

22

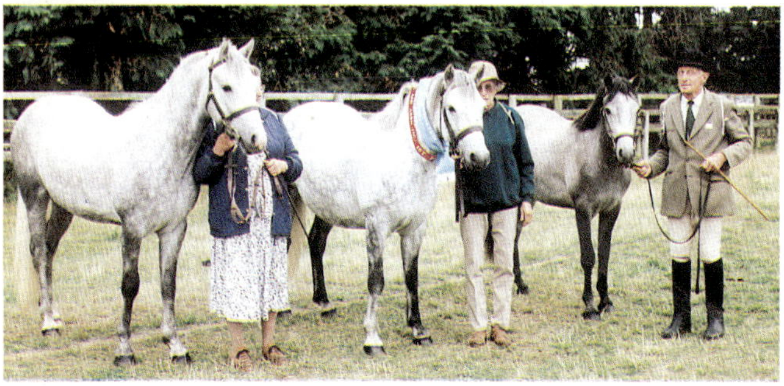
23

NEW ZEALAND

22. Canningvale Sunrise
23. Anne and Thady Ryan with Glenormiston Dingle Bay
24. Chiltons Way

AUSTRALIA

25. Tinderry Clancy
26. Ballintine Londonderry
27. Toorigal Danny Boy

25

26

24

27

to the next in search of a green bite somewhere and causing her owner annoyance in the process. Which parish may he find her in next? She bred Michael Hopkins three fillies by Inver Rebel[150] and a stallion son by Atlantic Storm. She then moved to Ballinasloe and Tom Kenny, who sent her to Paddy's Heaven four consecutive years. One of her daughters, Kinvara Hop, returned to the Hopkins family at Lettermuckoo.

Meanwhile one of her fillies by Inver Rebel was registered by Paddy Folan as Irish Harp. I remember seeing this mare at Maam Fair with a foal for sale. She was a small flea bitten grey, typical of the Inver Rebel stock. Paddy Folan was at the Gortmore inspection point in 1993 with a strong looking two year old grand-daughter of Irish Harp, Turlough Harp by Abbeyleix Owen. I caught up with another of Irish Harp's daughters in France, Camus Harp by Clonjoy. She was imported by Mr. and Mme. Boivin-Champeaux who farm in Normandy where she has bred them seventeen foals, all to Abbeyleix Apollo. One of her sons is at stud and a daughter, Safir de Beauchamp was Supreme Champion at Tours in 1991. What a fine daughter of Irish Harp and Clonjoy she has proved. After she moved to Ballinasloe, Grey Hop produced a filly foal by Mervyn Storm. She was named Grey Hop 2nd and appears to have only three registered daughters. The first Calla Lass was by Paddy's Heaven and went to Mr. Kamp in the Netherlands. The following two by Carna Bobby, were Coolfin Beauty and Gloves Misty. I knew there was an illustrious daughter lurking somewhere in the Grey Hop genealogy. For many years Gloves Misty graced the Clifden ring, often standing at the head of her class. Her good progeny have scattered but I was pleased to learn Eddie Madden has one of her daughters at Gloves, ready to take her dam's place.

In 1992 I drove along the Ballyconneely to Roundstone road making a few calls on the way. The first was to Martin O'Donnell of Lettershask and here I met his Killyreagh Kim mare, Swift Sally, with her colt foal by Smokey Duncan and her two year old filly by Mervyn Kingsmill. They were grazing a vast open expanse of tufty, old grassland which has its back to the Ballyconneely Bay. I spent an hour with my stud books that evening and only then did I realise, that in Swift Sally I had met a classic example of the essential staple stock of Connemara. Her family have been breeding at Lettershask for an uninterrupted fifty years. In 1940 John O'Donnell registered a Noble Star[151] mare and she was named Lettershask. Three of her daughters were registered, Evening Pearl by Silver Pearl, Brown Alice by Dun Lorenzo (who bred faithfully for the O'Donnell's for many years) and Lettershask Lady by MacDara who remained in Connemara but did not establish a family. Brown Alice had ten progeny registered between 1956 and 1969. Her third daughter, Careless Nora, by Clonkeehan Auratum, was retained at Lettershask and she was the dam of Swift Sally. It took me a while to realise that a full

[150] O.O.M. (See Index)

[151] S.I.M. 163 + 164

sister to Careless Nora, Wistful Alice was an old friend of mine. In 1968 my contact in Amsterdam[152] gave me an order for two registered mares. I had to travel to find them and answered an advertisement in Horse and Hound. I drove to Ross-on-Wye to see the mares Wistful Alice and Gold Ring. They both had golden coats and flaxen manes, and more important, Irish papers. I took them to my Buckinghamshire stud and prepared them for the journey to Holland. Some time later Wistful Alice joined Chiltern Arctic Venture[153] and the De Rooys at Brummen. In 1992 I took "Wistlys" photograph. She had reached a venerable thirty years and her back had gone but she was in good heart. She had, I reminded myself, been reared at Lettershask. Careless Nora sent her first filly, Calla Lass, by Paddy's Heaven to Holland and her second, Atlantic Noreen, by Carna Dun travelled to Sweden with Cregg Hero. She bred three by Rebel Wind and they were sure to be good ones. I hope they may not be too far from their birth place.

A good bit along the same road lies Letterdyfe which faces the Roundstone harbour and the Island of Inishnee, Letterdyfe "a hole in the hillside" is thought to refer to a small cave on the sea shore. It is here Michael Conneely lives and breeds Connemara ponies. I expressed the hope in Out of the Mist that Errisbeg Dolly might still be breeding in the region. On inspection day 1992 I mingled with the ponies who were congregated on the fore-front of the Roundstone harbour road and my eye was immediately taken by a steel grey filly. She looked well and showed quality. I discovered she was a daughter of Errisbeg Dolly and Mervyn Kingsmill and I talked with her owner, Michael Conneely. I asked whether I might see Errisbeg Dolly herself and he told me she was grazing at Aillebrack. I drove out there the following Sunday morning and called on Mrs. Paddy Macnamara for directions to her field. This visit revived memories of one I had made ten or more years earlier to talk with Paddy Mac about his heroic race pony, Battleship.[154] I soon had a car load of young Macnamara grand-children who were pleased to show me where I would find Dolly. This sixteen year old daughter of Abbeyleix Owen and Errisbeg Rose[155] fulfilled all my expectations. She was possessive of her filly foal by Smokey Duncan and did not allow me very close, but her strength of bone and build were easy to discern from a distance. I owe another apology, for the photograph said to be of Errisbeg Rose in Out of the Mist is, I am told, Zet Dixie. I'm including one of Rose herself to put the record straight. These two mares are not unalike in both type and build.

One day I was asked why Clifden Molly had not received a mention in either of my books. My first reaction was, why should she have? Then a glimmer of light dawned. Perhaps she was the dam of Abbeyleix Molly, the darling of the Clifden show and the Abbeyleix stud in the late sixties and seventies. I decided it was time to unravel the life and times of Clifden Molly.

[152] O.O.M. 201 [153] O.O.M. 201 [154] S.I.M. 80 [155] O.O.M. 282 + 283

In 1959 Colman Keane of Maam took his mare Knock Molly and her filly foal by Cilciarin[156] to Clifden show. It is more than likely the three of them walked the eighteen miles on foot. The mare and foal were both unplaced but by the end of the day had changed hands and were the property of John Mannion. He named the foal Clifden Molly and she spent the whole of her breeding years with the Mannions.

Mannion is a familiar name in Clifden and appears over many a shop window or main street bar. The family have had a long association with the Connemara pony from one generation to the next. Clifden Molly bred them seven registered fillies, four of whom were by Carna Dun with whom she nicked so well. It was the first of them, Abbeyleix Molly, who caught the judges eye as a two year old in 1967 and who went on to win her class in 1969 and again in 1972. She soon proved her value as a brood mare, providing more than one good daughter for the winners line up at Clifden. Two more of Clifden Molly's Carna Dun daughters passed through Abbeyleix and one of them, Millie, later came to England. As long ago as 1976 I judged Millie at the E.C.P.S. Ridden show where she gave me a great ride. At the time she was owned by Mrs. Betsworth, a sporting middle-aged lady with a very stiff leg who always mounted on the off-side from an orange box! Millie ensured she was completely safe at all times. In the spring of 1993 Millie is living happily at the Tyan stud in Wiltshire where she has produced a colt foal in her twenty-fourth year.

Clifden Molly spent her retirement years helping out with the Pony Club at Errislannan. She left John M (as John Mannion is known by his many friends) a daughter by Murphy Rebel, Gurtroomagh Molly. Young John P. Mannion takes riding lessons at Errislannan and is hoping her gelding son, Clifden Spirit, will be his very own riding pony one day soon.

In August 1984 I received a letter from Kathleen Maloney of Glann, Oughterard in which she shared her memories of Clifden show in the nineteen thirties. She enclosed a photograph of her Clonkeehan Auratum mare, Charm Bracelet, who was bred at Aughnaure by Paddy Joyce.[157] She bred Paddy four foals before being sold to Kathleen and by 1984 she had reared a further eleven, seven were fillies, all bar one by Jimmy's Fireball. A letter in 1992 told me she died three years previously, a few days after giving birth to her seventeenth foal. Two of her sons are registered sires, Woodquay Fireman in France and Glann Fireball in Co. Longford. In 1979 her yearling filly won a strong class at Clifden and a daughter, Kits Bay, has been retained at Glann to take her dam's place. Charm Bracelet was a gallant old lady who had the best of both worlds, spending the summer and autumn on the mountain with her foal and the winter months behind Kathleen's house with an open shelter and hand feeding. It was good to catch up with

[156] S.I.M. 159

[157] S.I.M. 107

another tough old Auratum mare, who was surely, in much the same mould as Muffy.

I was anxious to close this review of the Connemara Queens on a high note, and who better to do this than Smokey? In 1937 John Walsh of Kylesalia registered a mare by Connemara Boy,[158] Kilkerrin Cream. Her filly by Gil foaled in 1944 was registered by John Whelan, a chemist in Galway, and was given the name of Smokey. By all accounts she was a very sporting pony who looked equally well in harness or under saddle. The Whelans made the most of her talents and she was a familiar figure in the winners spot of the show jumping rings on the South Connemara show circuit. She bred the Whelans one foal, a gelding by Naseel,[159] who followed in the footsteps of his dam and was an outstanding performer. His photograph appears in Volume X of the C.P.B.S. stud book.

In mid-life, Smokey was given to Murty McGrath of Cregmore where she bred him a great line of ponies. The first, a colt by Doon Paddy, Thundercloud, was gelded and carried young Charles Hemphill boldly across the Blazers countryside. He was followed by two more geldings, the renowned performer Patsy Fagan by Tooreen Ross and a blue eyed cream, Tulira, by Clonjoy who proved to be an equally bold pony. He was sold to the U.S.A. where he carried Mrs. Onasis to hounds. Patsy Fagan was seven times Junior Event Champion in Ireland and travelled to Stoneleigh* as a member of the Galway Pony Club Team three years running. When following hounds there was nothing to stop him.

One of Smokey's sons was kept entire, Cregmore Dun, by Tooreen Ross. In 1963 Smokey produced her first filly, Our Smokey by Aura Dun and in 1966 her second by Bridge Boy, Lucky Smokey, foaled in her twenty-second year. Smokey spent the last years of her life in happy retirement at Tulira, where she died in the field with her son, Patsy Fagan, keeping guard over her.

Our Smokey was well hunted by Angela Hemphill before returning to Cregmore to take the place of her dam. She bred Murty three colts, Lakeview Rambler, Cregmore Colm and Smokey Playboy. Colm went to Australia and Playboy has taken the place of his sire, Tulira Mairtin,[160] at the Tulira stud, Raford. Our Smokey bred four fillies, two by Atlantic Storm, who were not retained, and two by Kilgreaney Lad, Smokey Jane Grey and Smokey Latecomer, the mares you will find in the Cregmore paddocks if you visit today. Lucky Smokey was more frugal in her breeding habits and has only four registered progeny, two of them being the stallions Cregmore Kiaran and Smokey Shane.[161] The latter displayed plenty of his grand-mother's jumping talent for the Cheviot stud in Scotland.

Cregmore Colm by Tulira Rambler was used by the Tulira stud until he was exported to Australia in 1981 where he spent ten years with Mr. Parsonage of

[158] S.I.M. 167 [159] S.I.M. 196 + 197 * B.H.S. Headquarters, Warwickshire [160] O.O.M. (See Index) [161] O.O.M. 50 + 51

Wilberforce, New South Wales. He was used primarily as a teaser and rarely had the opportunity to cover Connemara mares. Ten years later he made the long 700 mile journey by road to Penny Brown's Celtic Stud at Ararat in Victoria. And so in his twentieth year Colm has been given a second opportunity to prove himself a good sire of pure bred stock. Penny already has a useful crop of yearlings and foals by him and looks forward to their progress.

Before Colm left for Australia he sired a number of fillies and one very good son, Tulira Crackerjack,[162] so named because there were 'fire lighters' in the Ardrachan area when he was registered. Crackerjack went to Mr. Paul in France where his courage, in spite of his lack of inches, soon brought him to the attention of those in the competition rings of show jumping and dressage. In 1992 he won the French National Dressage Championship and has since been retired.

When I wrote Out of the Mist I was unfamiliar with the Smokey family but was happy to accept Graham Tulloch's view, "it was one of the best". Graham had recently purchased Smokey Duncan, a son of Smokey Jane Grey and Westside Frank. Duncan was bequeathed to the Society on Graham's death and won the Clifden Championship in the hands of Josie Conroy in 1990. Duncan has matured into a strong pony and was one of the stallions chosen for Performance Testing in 1992. His trainer's report read; "A super armchair ride who would take on any cross country or show jumping fence". As a great grandson of Smokey you would expect no less.

[162] O.O.M. 244

The Stagg family at Clifden 1991

The National Park Ponies below Diamond Hill, Letterlrack.

CHAPTER FOUR

LOOSE ENDS

Readers may find these ends very loose. They are a dip into a fountain of memories of people and ponies. And an expression of thanks to some of those who have helped me, written to me, and encouraged me on my way.

Toon Duinhof, was a Dutchman, who enjoyed a love of horses and ponies from an early age. His English wife, Edith, is of Irish descent, and for this reason they decided to visit a Connemara pony show in Holland in the nineteen-sixties. Toon was immediately attracted to the breed and subsequently bought three Connemara ponies. In 1976 Toon and Edith visited Ireland for the first time and, while in the West, Toon bought himself a copy of the Connemara pony stud book. This aroused his interest and he began a detailed study of the various blood lines. It became a painstaking task identifying the progeny of every sire and every dam registered in the Irish stud books.

I found the result of some of his work in Garnet's house in 1980. I wrote and asked him if he would write an article for The Chronicle to show us what he had learnt. His response was "Stud Books - How do we use them" which appeared in Volume 8, 1981. From that time on Toon and I were regular correspondents, and he generously responded to my requests for information on this sire or that mare.

When I was in Holland in 1986 I visited the Duinhofs at their home, where Toon showed me his filing cabinet which contained all his records. He was, he said, happy to share them with me at any time in order to assist my writing. When Toon died in 1990 I asked Edith whether she would consider bequeathing me this valuable information. She and her sons agreed to my request and it is now carefully filed under "Duinhof Data" at Combe Cottage. I turn its pages again and again and know I shall always do so.

Toon was a business consultant by profession and no doubt it was his methodical mind which enabled him to undertake such a task. He did not use a computer. All the stud book numbers were recorded in his own neat hand. It was a labour of love for which those interested in the breed will always be grateful, no one more so than myself.

A mile or so beyond Moycullen there is a hidden left hand turning which leads up the mountain road to Pillagh. A right hand turn at the T junction at the top brings you to the O'Connor homestead where Joseph was reared and where he and his family live to this day. It was his father, Matt, Joe told me, who had an interest in the Connemara pony and it was he who bred and worked the Pillagh ponies between 1955 and 1985. A grand-daughter of Pillagh Beauty, Eudene Moonstone, led me to this particular mountain top. I judged her as a two year old at the Surrey County show in 1991. After placing her Reserve Champion I enquired her breeding and later met her dam, Pillagh Grey Pearl, at the English breed show. I noticed Marble had left his own particular mark on this flea bitten, strong boned, honest looking daughter of his. The first pony registered with the Pillagh prefix, Pillagh Star, went to Mr. Wright in the U.S.A. Did she leave the second one, Pillagh Beauty by Cilciarain[163] foaled in 1955 behind her? (Beauty has no dam recorded in the stud book). Pillagh Grey Pearl was the eleventh and last of Beauty's registered progeny. It was Matt O'Connor's habit to walk Beauty across the mountain to Knockranny each year, to visit whichever stallion Peter Kyne had in his care. How was it, I wondered, that Pillagh Grey Pearl was by Marble who lived in Galway? Joe explained how his father sold John Brennan one of Beauty's foals and a free service to Marble was part of the deal, a valuable one in the year of 1973 when Beauty visited him. Grey Pearl reared at least six foals on the Pillagh mountain range before being shipped to England, where she was discovered by Ros Harris and Anne Westhall in a pitiable condition. A rescue plan was put into operation but it took some detective work to establish the sire of the foal she was carrying, Eudene Moonstone by Ardravinna Andy.

As a boy, Joe remembers leaving home at 1.00 a.m. in the morning to help his father Matt walk the sheep and the cattle to Galway market. Pillagh Beauty accompanied them drawing a turf-cart piled high with turf, a valuable source of income for the O'Connor family. Her foals were taken to Spiddal Fair to be sold in the autumn. Today the risk of ponies straying off the mountain onto the busy road below has increased. Trade for foals at Spiddal Fair is slow and the turf travels by truck to the purchaser. Joe's good looking son, Rowan, is studying for a Degree at Galway University and will never experience the joys and hardships of rising in the small hours of the morning to walk the road from Pillagh to Galway. It seems likely the Pillagh prefix will disappear from the stud book but this hardy strain of ponies continues to thrive to the South and East of Galway as well as in Surrey and elsewhere in the world.

Another of Marble's daughter's has long been a favourite of mine. In 1970 she was my choice to win the eighty-four strong entry for the yearling filly class at Clifden. In the event she, Celebration Cream,[164] was placed fourth. In 1973 I met her again in Mr. Healy's fields at Salthill. She was only four years old but already had a yearling colt and a colt foal on the ground. It was a grave temptation to delude myself the yearling would fulfill the needs of the Chiltern Stud, so much did I admire his dam. He was by Finney Master[165] and altogether too well grown for a yearling. I made myself look the other way. Napolean, as he was later regis-

[163] S.I.M. 159 [164] O.O.M. 263 [165] O.O.M. 224-226

tered (how, I wonder, did he escape being called Napolean Brandy?!) spent a short while at Tulira before being exported to Libya.

In 1974 Denis and Anne Wilson of the Cheviot Stud were making their first visit to Connemara in search of a good mare. They asked for some guidance about where to look and I suggested they try to buy Celebration Cream. If you covet a mare and cannot own her yourself, what better than to see her in a friend's paddock? Celebration arrived at her new home on the Scottish Borders carrying a foal to Windy Cove Ranger,[166] which Anne was unlucky to lose to grass sickness in 1980. Celebration's record as a brood mare could hardly be bettered. Two of her fillies, Cheviot Bryony (Windy Cove Ranger) and Cheviot Clover (Millfields Commodore)[167] won the E.C.P.S. Youngstock Championship in 1977 and 1978. Cheviot Crocus (Island Duke)[168] was Champion foal in 1979. Her first filly by Smokey Shane grew over height but there are high hopes for her full sister, Cheviot Champion, who has recently joined the Hayselden stud in Kent.

Celebration bred six colts in Scotland, one by Chiltern Humdinger and five by Smokey Shane.[169] Cheviot Midas went to Miss Held's Stud in Switzerland where he is her successful foundation sire. Cheviot Hemlock spent some months at the Chiltern Stud in the hope he might take Island Duke's place one day. He was a very precocious colt who looked like growing over height and we did not persevere with him. The geldings have all made excellent dual purpose ponies and Cheviot Bracken won the E.C.P.S. Performance Award Scheme in 1990.

My delight in these two Marble mares increased when I discovered they shared a maternal grand-sire in Cilcairin. They typify so well all that is best in the breed, unalike in stature and quality, maybe, but unmistakably Connemara ponies.

Whenever I drive past the recently restored Glenlo Abbey* and its surrounding newly designed golf course, my thoughts drift back to Carna Dun and the seven semi-retirement years he spent at Glenlo. In 1968 a consignment of Connemara ponies was sold for export by Johnny Palmer of Glenlo Abbey. The deal was subject to them being certified in foal. Banba Chestnut, by Island King,[170] was one of them. She failed her pregnancy test to Carna Dun and was sold, as barren, to Graham Tulloch. The following spring she unexpectedly produced a fine filly foal who was named Ocean Bambi and sold to Colonel Tony Morris of Ballinaboy. Bambi was one of two Carna Dun mares still alive when I was in Connemara in 1992. She has visited a wide variety of Connemara sires in her busy breeding life; Kimble, Gold Fort, Sarsfield, Abbeyleix Owen, Loobeen Larry and in 1992, aged twenty-four, she was visiting Westside Fred. One of her daughters, Ballinaboy Jonquil, is a regular breeder for the Templebready stud in Co. Cork. A son, Ballinaboy Eamonn by Larry is foundation sire for the Tir na n'Og Ranch in

[166] O.O.M. 69 + 236
[167] O.O.M. 141-142
[168] O.O.M. 53-54
[169] O.O.M. 50-51
* Glenlo Abbey between Galway & Moycullen
[170] O.O.M. 136-139

Florida. A full sister to Eamonn remains at Ballinaboy to take the place of her dam one day.

Two years later, Banba Chestnut bred Graham a colt to Doon Paddy, Ocean Seames. He travelled, with his dam (who was carrying a foal to Kimble) to the Arenbosch stud in Holland before moving on to Germany, where he has left some useful brood mares. The Kimble filly, Arenbosch Frederique won the Dutch show Championship in 1986. I tend to think of Australia as 'Island King Country' and was interested to discover such a useful daughter as Banba Chestnut leaving her mark in Connemara and elsewhere.

Silver Lass and Tully Blue[171] arrived in England in November 1971. I drove to Birkenhead to collect them. Before long they moved on to new homes, Silver Lass to a friend near Aylesbury and Tully Blue to the Hobbans stud. Silver Lass had an unusual pedigree. Her sire, Errigal Prince, was not highly regarded until he moved to South Connemara where he gained fame as a jumping pony. The dam of Errigal Prince, Errigal Princess, was by Tiger Gill[172] who was known to jump out of anywhere if the mood so took him. The dam of Princess, Furbough Lass, was by the Irish Draught sire Skibereen.[173] A glance way back into the past at Lt. Col. Royston Piggot's mare, Boffin, provides further evidence of ample jumping ability. Florence Nightingale, the dam of Silver Lass, was by Cilcairin and he too was renowned for jumping from one enclosure to another. What a wealth of jumping talent in the pedigree of one rather ordinary little mare.

Silver Lass produced a filly foal to Little Joe in 1972 and visited Island Duke with the filly at foot. I wonder what became of her, Woodham Silver Pearl? In 1973 Silver Lass produced a colt to Island Duke. He was gelded and named Silver Duke and as a three year old he went to Dr. Nicola Hall. He was her first pony and she, so she tells me, was an adult novice. While Nicola was completing her training as a Doctor, she and Duke travelled from one part of England to another. They hunted with a variety of packs of foxhounds, always taking the lead when given half a chance. They tackled the South and West Wilts, the Suffolk, the Devon and Somerset, and the Vale of Aylesbury country-sides. While out with the latter pack, Duke caught the eye of Phil Oliver, father of Alan*, who would dearly have liked the opportunity to train him as a junior show jumper. Nothing could tempt Nicola to part with him, and together they have enjoyed seventeen years, Fun with a capital "F". They became popular and regular contestants at the E.C.P.S. ridden shows and Duke's name is engraved on the Emmet, Poigin, Wisbridge, Hellings and Freddy Grant Cups. He has often held a place in the Performance Award Scheme results and has awarded his own 'Silver Duke' Trophy for the highest placed Adult Rider/Driver.

[171] O.O.M. 23 [172] S.I.M. 158-159 [173] S.I.M. 193

* Alan Oliver, International Showjumper

I have not given Island Duke any credit for the exploits of his son but I would hazard a guess that his natural gaiety and zest for all that he does are the rich gift of his sire. In recent years Silver Duke has had to compete for Nicola's time while she produced a family. I hope she will find one as good as he on which to mount them when they grow up.

One of the compensations for the labour and risks involved in writing and publishing your own book is to receive readers' letters of appreciation. Sometimes these letters contain news of ponies which fill a gap in my research. Other times they simply say thank you and include tidings of their own special pony. Photographs are always welcome and occasionally readers accept my invitation to tell me where errors have occurred in the text. Isabelle Bouyssi wrote from her Aubier Stud in Charentes, which she says is Cognac country. In Out of the Mist I mentioned a mare, Screen Lady,[174] and regretted I had no knowledge of her female progeny. Isabelle wrote to say she had a much loved daughter by Robber Boy[175] foaled in 1978, Milady de Ravary. Hubert Laurent's Melody stud has also had valuable service from Eva de Ravary, another of her daughters, by Island Earl.[176] How satisfactory to learn Screen Lady has provided French breeders with good brood mares.

Isabelle's herd includes mares which would make many of us envious. Among them are; Fort Sapphire by Glenarde out of Moycullen Mite (who is half sister to Celebration Cream. Moycullen Mite has left a good Marble daughter, Fort Hazel in Connemara where she is one of the Letterfrack Park herd); Ashfield Gypsy by Thunderbolt out of Orphan Dolly; Tulira Linnet by Carna Bobby; Easter Thorn by Carna Dun and Easter Queen by Island Duke. Her senior stallion is by Tulira Mairtin, Kid de Coislet, and she has recently imported an English bred son of Thunderbolt, Cocum Thunder Boy.

Hubert Laurent has kindly answered some of the queries raised in Out of the Mist. The sire, Windy Cove Ranger, was refused a licence after three years service for having grown over height. His good son, Idenoir,[177] has five licensed sons in France and so this line is in safe keeping. The photograph I included of Randam du Bailly[178] is not he, but Mr. Chupin's stallion Tynagh de Briace. Randam du Bailly has, in any event, been gelded after proving sterile.

France is said to have the greatest number of registered Connemara ponies in Europe. This I can believe, when I hear that Mr. de Pas alone has three hundred and fifty ponies in his Pony Club, sixty of which are Connemaras. No doubt it is significant that horse and pony breeding is generously subsidised by the State in France. A good (non-horsey) friend of mine, who is confined to a wheelchair, wrote to me after a visit in 1991 to the St. Lo Haras in Normandy. It was the high-

[174] O.O.M. 245
[175] O.O.M. 243
[176] O.O.M. 240
[177] O.O.M. 236-237
[178] O.O.M. 239

light of her holiday to find one hundred and thirty stallions from Thoroughbreds, to Percherons to the Connemara pony, each stabled in spacious, airy loose boxes. She understood the Haras provided winter quarters for the sires, the majority of whom were sent out in the summer months to wherever their services were required. For my benefit, she took a photograph of Indra Rebel behind his stable door. In 1981 Indra Rebel won the stallion class and the progeny group at the French Society show held at Cluny. (His photograph appears in Volume 9 of the Connemara Chronicle). He is not registered in the Irish stud book but it is likely he was bred in the mid 1970s.

A postcard from Francois Birkigt, another dedicated French breeder from the Le Chesnay stud wrote, "Your book is super. Cliff (Knockadreen Holiday Cliff)[179] is very happy you call him King. He says you are the only one who realises how marvellous he is".

After receiving her copy of Out of the Mist an Australian breeder wrote "To hell with the house-work. I can't put it down". Another kindly said "I read it like a novel first and now it is a valuable reference". She went on to point out that 'Ranch' was an incorrect term when referring to land in Australia. It is, she tells me, either a 'property' or a 'station'. And I apologise for using the word 'rustling' instead of 'mustering' in the same paragraph. Of course Margaret Campbell and Inis O'Hara[180] were not stealing their own cattle!! An enthusiastic new recruit to the breed in Australia, is Mark Osborne of Geelong, Victoria. He owns a half share in Wychwood Grey Gull, a daughter of Coomel Kestrel[181] and Ardan.[182] Mark was soon bitten by the Connemara pony 'breeders bug' and together with his wife and two small children made a 3,500 kilometre round trip to visit Ardan in Kanmantoo, South Australia. He also spends time studying the Australian stud book, 'travelling through the mists' and visiting stallions most local to him, in order to decide on a husband for Grey Gull. Good luck to him and others like him.

In June 1992 Ardan celebrated her thirtieth birthday with a Barbecue lunch attended by many Connemara admirers. She joined the Wychwood Stud in 1975 and her death at Christmas in 1992 has left a big gap. Ardan was a truly great 'little un' being no more than 12 hands 2 inches. She would be in danger of being disregarded in Connemara today but Australia will be forever grateful to this little mare for her unique contribution to their stud book.

Jack and Heather Storey and their daughters Kate and Rosemary have long been loyal supporters of the breed at their Blandings Stud at Arthurs Creek. Heather's scholarly response to and reviews of, my books have been much appreciated. Visitors to Arthurs Creek are always welcome, as Bill Concannon discovered

[179] O.O.M. 242 [180] O.O.M. 144 [181] O.O.M. 144
[182] O.O.M. 146-147

recently. Bill is a Melbourne business man who, after twenty-five years in Australia, has an Australian accent and an Australian passport. He is, however, a born and bred Galway man whose native home is Truskey East, Barna. In the early nineteen-sixties Bill owned a little mare, Furbough Heather, who was by Cilciarain. She visited Clonjoy, the nearest sire to home, and year by year produced nondescript colt foals. Bill was persuaded to part with her, to a friend in Aughrim, in order that she might visit a different sire.

I cannot drive by Aughrim, nor indeed mention its name here, without pictures flashing through my mind of the fearful battle waged there on the 12th of July 1691. It was a defeat for the great Patrick Sarsfield, but an honourable one, most believe. Piers Wauchape writes "The earth shook from the pounding of thousands of hooves. Hundreds of pistols were discharged at close quarters before the troopers drew their swords and stood up in their stirrups to charge, roaring their lungs out". It was a bitter battle and the greatest slaughter in Irish history. Sarsfield later wrote:

"This useless sword shall open every vein,
And steep with Sarsfields blood fair Aughrims plain"

It was on "fair Aughrims plain" that Furbough Heather produced a filly foal to Paddy's Heaven in 1966 (not for the first time, I am frustrated at my lack of any knowledge of Paddy's Heaven). Bill retained an interest in the filly, Barna Honey, and produced her to win the yearling filly class and Clonkeehan Cup at Clifden in 1967. Honey was sold to Mary McCann of Co. Meath and returned to Clifden in 1972. She, and her foal by Checkpoint Charlie, were both placed third.

Bill arrived at Arthurs Creek to see the Storey ponies, with a photograph of Barna Honey in his hand. He left with a copy of Shrouded in Mist and sat up through the night to read it from cover to cover. On a recent flying visit to London and Galway he got in touch. We had two telephone conversations swopping memories of mutual old friends, both human and equine. To chat with Bill brought a little sunlight into two drab March evenings. He told me Shrouded in Mist had made him homesick, the nicest compliment he could have paid me.

Apologies to New Zealand for mispelling many of the place names in their chapter. And to Winnie Tatham also. She took the photograph of Lonsdale Windfall[183] and it is therefore a friend holding Windfall and not Winnie herself. Chilterns Way does not have an 'own' to his name. I cannot think where I picked that up! He has been re-registered as Chiltons Way at the request of the keeper of the stud book.

Caroline Nesbit in her review of Out of the Mist for 'The American Connemara' writes, "obviously mistakes are going to creep into any volume of this

[183] O.O.M. 159

scope ..... I don't remember running across any that would merit 'the major blooper' category". That was kind of her. One or two errors in the U.S.A. chapter have surfaced which do concern me. I received a delightful letter from eighty-four year old Dorothy Lyons in California who "loved me madly for all the nice things I said about her people", her people being her pony family. I was distressed to find I had confined her great pony sire, Bar Bar A's Kilkerrin Paddy[184] to 'greener pastures' when he is, in reality, enjoying happy retirement as a gelding. Dorothy tells me she has a 'yen' to travel by Concorde to the U.K. and meet up with all her old friends in Clifden. I hope I have half her spirit, should I reach the age of eighty-four.

A generous appreciation from Ed and Jackie Harris noted I had omitted their prefix, Hideaways, from their home-bred sire Greystone Alex. This was an oversight I regret. As a breeder, I know the satisfaction you feel when you have bred a good stallion. Jackie's letter concludes, "Don't forget to include Hideaway Farm in your itinerary when next you come to the U.S.A. For now, our very best and know that we think of your work with gratitude".

My good friend Catherine Mack who lives in Middleburg, Virginia was in bed nursing a bad back when Out of the Mist arrived on her door mat. What a happy coincidence. She told me it acted as "a total opiate and natural healer" and gave her courage to get off her sick bed and make her breeding plans for the spring.

Nina and Anders Gustafson, long standing breeders in Sweden, make regular visits to Connemara and England. Nina, together with Jan Jango's bookshop in Johanneshov, have given valuable help with the distribution of my books in Sweden. Recently, the Gustafsons were sad to lose their twenty-seven year old foundation mare, Gina of Ganty. Nina remarked how successful a riding pony she had been, but an unpredictable brood mare. She referred to John Brennan's[185] comment about her female ancestors 'who produced more variation in colour and type than any other'. Nina's desire to breed the best has encouraged her to take a mare to Denmark to visit the stallion Hazy Dawn.[186] Enterprise such as this is sure to reap rewards.

I have no more regular correspondents than the Milleders in Germany. Their enthusiasm for the breed encourages Edith to undertake all the secretarial work for the Society, and Beatrice to spend numerous hours compiling a record of the breeding stock in Germany. So many of them have been lost, because Germany has no central register or stud book. Their appreciation of my work is expressed in many ways and I welcomed a letter from Beatrice questioning some of my observations in the German chapter. It stimulated an exchange of views between us.

[184] O.O.M. 100-101 [185] O.O.M. 42 (Irene Grey) [186] O.O.M. 189-190

Every letter received from Gerda Merrild of Aeroskobing in Denmark is a treasure. Gerda's feelings go very deep, and expressed in her broken English it is easy to share them with her. When Bjarne's work, as a surgeon, took him to Lapland the Merrild's Nicholas[187] herd of Connemara ponies had to be dispersed. This has never ceased to be a source of sorrow to them. Out of the Mist stirred many old and happy memories for Gerda. She wrote; "Closing my eyes I see all our dear Connemaras, our near family, and I hear them. And a spring without affectionate foals and going and coming without the ponies standing at the gate looking after us, neighing, is bad. In old days I was so silly. I waved my hand to them and shouted 'I am returning soon'. Not to have a Connemara putting the head on my shoulder and not able to put my nose into the hot coat on a sunny day and enjoy that smell. That is bad, bad."

After reading Out of the Mist, Joan Dolan of Kilternan, Co. Dublin was prompted to buy her first Connemara pony. Only the best would do and the two year old filly, Moorland Juliet, winner of the two year old class in 1991 was bought at the R.D.S. I am sure she will be expecting her first foal any time now.

Please turn to page fifty-two in Out of the Mist and note that Midnite Marcus is by Rustle of Spring and not his grand-sire Celtic Truska Prince. And a similar error is made with regard to Millfields Ensign on page two hundred and nineteen. He is, of course, by Wisbridge Erinmore. Thank you, Nicholas Palmer, for highlighting these discrepancies on the back of a beautiful postcard of Desert Orchid, my favourite race horse. A further grandson of Macnamara, Nyton Village Master should have received a mention on page thirty-six.

On the tenth of July 1989 as I walked out of the Presteigne Chapel I was greeted by a lady with an American accent who was looking for Pat Lyne. It was Judy Brown-Bieber with her husband Dale and two of their children, Kirsten and Benjamin. They were passing through England on their return to the U.S.A. Dale, a Doctor, had spent two years working in an African Mission, a project sponsored by his practice in the States. While there, Judy read Shrouded in Mist. Her visit to Presteigne was to share her enjoyment of it with me. It was a spontaneous, unplanned visit, the memory of which I treasure.

An equally rewarding visit was made by Doris Davidson of Kelowna, British Columbia. She drove half across England with her cousin Anne Dain in order to meet Arctic Moon and myself. It was during the last month or two of Moon's life. Doris wrote after her visit, "My stomach tightens when I think of the gallant old mares who can go on no longer". How well I know that feeling. She enclosed a photograph of her three month old colt whom she had named "Sky Lyne" and apologised for not asking me first. She need not have feared I would mind.

[187] O.O.M. 184-186

As a seven year old, I prized a small autograph book in which my grandfather Lyne signed his name beneath a quotation "I freely admit the best of my fun I owe it to Horse and Hound..Whyte Melville". It was many years before I learnt to share my grandfather's devotion to his weekly copy of Horse and Hound. In 1970 Horse and Hound carried a report on the Clifden show, which was unusual. It appeared under the bold heading "A CONFUSION OF CONNEMARAS" and the contributor was Dalesman. Bay De Courcy-Parry rejoiced in the non-de-plume of Dalesman, and was a regular hunting correspondent for Horse and Hound. In 1943 he was said to have provided "a new injection of descriptive writing and humour which was badly needed". Throughout his life Dalesman hunted several different packs of hounds, wherever the opportunity presented itself. He did not appear to stay in any one place for very long and I have no idea how he came to be in Westport, Co. Mayo in the nineteen seventies. In 1975 I was looking for someone to report on Clifden show for the Chronicle and I wrote to Dalesman asking if he would kindly do so. It was only the third edition of our unsophisticated journal, and twenty years later I wonder how I had the audacity to ask someone with a reputation such as his. His reply, an article "A little knowledge is a dangerous thing" and a page full of sketches have sat in my archive file ever since. The article was a cleverly written 'skit' on the pony show and Connemara life in general. I did not consider it suitable for publication in the Chronicle but I am pleased to have the opportunity to include his letter and self-portrait here. They are a valuable memento of an interesting man who responded generously to my request ..... with no payment asked.

Telegrams were an important every day method of communication in Connemara, and were transmitted from one P.O. to the next, from Carna to Roundstone, to Ballinafad, to Clifden, to Moyard. Each of the post-offices operated a manual telephone exchange and it was the post-masters duty to deliver the telegram. Willie O'Malley did so on his bicycle, receiving no more than a shilling a time for the service.

In May 1992 I drove along the L 102 link road between Roundstone and Ballinafad in warm spring sunshine. I passed through Toombeola and by the dancing waterfalls and fishing pools which spill out of the nearby Ballinahinch lake. As I approached the old Ballinafad P.O. I noticed the phone box by the road side and a lady working in her garden nearby. Memories of my visit in 1978 flooded back and I recalled how Willie's[188] long memory provided a valuable paragraph for Shrouded in Mist. I took the opportunity to say Hello to Mrs. O'Malley, and together we shared some old memories. I learnt how Willie had died on Christmas Day a few years back. Ann had been reluctant to cook a turkey for just the two of them, but Willie insisted it would not be Christmas without. As they sat down to lunch there

[188] S.I.M. 111

was a knock on the front door and they found an uninvited guest on their doorstep, a friend who lived on his own, away off in the heart of the mountains. In order not to offend either of his nearest neighbours, by joining one of them rather than the other, he made the long walk to the O'Malleys. He was an Angel of the Lord in disguise. They enjoyed lunch and each others company and later as Willie relaxed in his armchair, he died. How would it have been for Ann without her uninvited guest?

The passing of the manual telephone exchange marks a significant change. The personal touch has gone. So often there was no need for your call to be plugged in, "Ah sure, Michael will be over at his mothers and will not be back till late" or "there is a meeting, you'll be finding no one at home tonight" or "he is over visiting Josie, will I put you through to Josie's?" Yes, that is how it was. Today automation has taken over. Taken over it may have, taken the place of, it never can.

I first visited Pat Mulkerrin (Tom)[189] and his wife at Callowfeenish in 1976 and have returned on at least three occasions. I never cease to be amazed at the resilience of this remarkable couple. There was no question of taking life easy once they had reared their family. A two storey house, suitable for guests, was a long held ambition. Mrs. Mulkerrins followed her sons to the U.S.A. where she kept house for them, going out at night to earn money nursing the sick or washing up at a restaurant. The proceeds were sent home each week to Pat Tom, who bought the materials required for the building, which he undertook himself. The day came when the task was completed and Mrs. Mulkerrins returned to Connemara to open the door to her guests. Many of them were students, and it was like having a big, happy family all over again. Twenty years on, the Mulkerrins have grown old, and are no longer able to climb the stairs of their two storey house or to welcome guests. One of their sons has returned from America and farms the land with modern machinery. The days of Carna Dolly and her cart over-flowing with sea-weed or turf are a hazy, precious dream. Pat Tom's greeting when I visited was invariably a glass of his own special brew, whiskey laced with lucozade. I am unlikely to forget the Mulkerrin's own personal brew, of stoicism and endeavour.

The Derrynea inspection point invariably draws an interesting selection of ponies. They walk in from Crampaun, Derrykyle, Glenicmurrin and from as far off as Lettermullan and the Island of Inishbarra. Sean O'Flaherty of Carraroe was there in 1991 with an attractive grey filly. She was, he told me, by Sarsfield out of Dimple and he named her Carra Dinnish (Honest Friend). Some time later, I delved into her pedigree and discovered Dimple was descended from the same female source as my old friend, Leam Lassie. Dimple's dam, Leam Darling

[189] S.I.M. 207-209

(Bridge Boy) was out of Irish Darling (Dun Lorenzo) and she was a half sister to Leam Lassie. They were both out of Irish Lass, whose dam, Bog Oak, was foaled in 1930 and registered number 287. Bog Oak[190] was almost certainly a descendant of the one and only Cannon Ball.[191] On the other side of the coin, Dimple's sire was Clonjoy and I do not doubt it was his influence which drew me to Cara Dinnish. I hankered after the filly for some time, only to discover she was not for sale. None the less, I determined to find Dimple one day.

A left hand turn at the 'Póitchin Still' at Inveran, on the Casla to Spiddal road, leads to Aille, the home of James Carr. As I drove along the approach track, I realised I was driving towards the mountain range on which Malcolm Wallace ran his herd with Mountain Lad[192] and Tully Lad early this century. They were as free and wild as the wind, but of an incomparable hardiness. This is where Dimple lives with her two year old filly by Beal Tra and her yearling filly by Sarsfield. James Carr fetched them off the mountain for me to see on the strict understanding they were not for sale. Dimple is a chocolate dun in her twenty-first year and knee high to a grasshopper. Yes, I found her to be in the mould of her family, and no more than the 12 hands 1 inch which the stud book tells us. Her tail had been chewed off, her coat was rough and she was carrying another foal. She was not short of condition and James held her by her forelock to show me the large dent in her forehead, no doubt the reason for her name. Dimple has the kind, full eye of her sire and her progeny are evidence of her value as a brood mare. She continues to draw the turf home for James and it was apparent they understood each other very well. As I drove away from Inveran, in a soft evening light, I mused how good is was to meet men like James Carr and Sean O'Flaherty who keep their ponies for pleasure rather than for profit. It is important these old and valuable blood lines continue to breed and to flourish on their native heath.

My visits to Ballinafad, Callowfeenish and Dimple bring me full circle and maybe they tie some of the loose ends together?

[190] O.O.M. 147 [191] S.I.M. 147-150 [192] S.I.M. 177-178

CHAPTER FIVE

TRAVELOG

I had many ambitions, but to travel was not one of them. My personal aspirations lay rather closer to home; to win the hack championship at the Royal Richmond Horse show; to train and ride the winner of the ladies race at our local point to point; to be an energetic secretary of the Whaddon Chase pony club; to play an active role in the parish to which I belonged. And most urgent of all, to earn enough to pay the bills at the end of each month! The first two ambitions were never fulfilled but the others, by and large, were. I was content with life as it was, in spite of the demanding work load it invariably entailed.

Horse-buying expeditions to Ireland in the nineteen-fifties were my first experience of travel, and for some years I stubbornly refused to fly. I regularly caught the boat train at Euston at eleven p.m., boarding the boat at Holyhead in the small hours of the morning, arriving at Dun Laoghaire at seven a.m. where George would be waiting for me. Eventually I was forced to discover how time-saving and simple it was to fly from Heathrow to Dublin and I overcame my resolve never to enter an aeroplane! By 1967 I was brave enough to join a team of 'Peregrini'* who volunteered to spend their vacations working in deprived areas of the U.S.A. We travelled very comfortably with Aer Lingus and from that time forward flying held no terrors for me. I made similar organised flights to Russia and Iceland in the nineteen-seventies. Little did I realise how many opportunities for travel the Connemara pony would provide in years to come.

How to encapsulate the memories of sixteen journeys, made to ten different countries, during a span of twenty-five years, occupied my mind for many days. I was aware a litany of dates, place names and champion ponies would be but an empty shell. As I relived my journeys, this Travelog became a sketch book of personalities, places and ponies, and of remarkable and unremarkable events. I hope it gives a glimpse into the life and use of the Connemara pony in foreign lands.

Those who read Out of the Mist may remember how my plans to deliver ponies to Holland in 1968[193] foundered. I had to wait until 1977, before a similar opportunity to travel ponies by sea presented itself. By this time Bent Nielson had established a flourishing herd of Connemara ponies at his Oxenholm stud in Denmark. His enthusiasm for the breed was such that he planned an International Festival for Connemara ponies at Roskilde. This was an entrepreneur at work and it required the skills and finance of such a person to organise an event of this scale. It was an exciting venture for those of us who became involved. The Cocum stud

* Travellers for Christ

[193] O.O.M. 201

undertook all the travel arrangements for the English team and we filled a horse box with eight ponies for the journey from Harwich, where we all met, to Esberg and thence to Roskilde. Benita Batchelor (now Sanders) and I, travelled with the ponies and met the Williams and Millers in Denmark. The ponies travelled well but picked up a virus - possibly at the overnight lairage stables - and arrived with temperatures and coughs. Immediate treatment reduced the temperatures but not the coughs, and five out of the eight were well below their best on show day. Fort Viola won her class, but the rest of us, the Rosenaharley ponies, Laurin and Rossleague, Cocum Jackpot, Washlands Boann, Wisbridge Golden Virginia, Tiercel Witch Hazel and Chiltern Cameo had to be content with lower placings.

The Festival, however, was not all about winning. It was about meeting breeders and ponies, and sharing our common interest. The Oxenholm stud took the major honours, but the Swedish team of sixteen ponies put up a brave show in the ridden events, winning the Dressage and Jumping classes. Four ponies from Germany and five from Ireland were entered but Kurt Hillnhutter and Margo Dean were their only competitors. Representatives, without ponies, from Australia, America, France, Holland, Spain, Norway and Finland were also present. It was a grand plan to bring us together and it certainly succeeded. The Danish hospitality, with a strong Irish flavour, was handsome. A popular Irish group were invited to entertain us with their music, and the evening meal was complimented by the liquor, Irish Mist.

Dan Llanigan O'Keefe, together with judges from Sweden and Denmark, undertook the task of sorting the two hundred ponies catalogued. The European exhibitors wore white trousers and coloured tops and we English looked very English in our tweed jackets and collars and ties! No one could argue with the Championship which was awarded to Marble,[194] with the elegant Danish bred mare Oxenholm Bardot in reserve.

On the return journey, the English competitors sunbathed on board deck, sharing a bottle of champagne with their President, the late Sheila Inderwick (who had judged the Dressage classes), and their new Australian friend, Tony Royal of the Hunters Lodge stud in New South Wales. We were celebrating 'Horses Birthday' held, by tradition, on August 1st every year in Australia.

Four years later, in 1981, Bent decided to run a similar Festival at the same venue. English breeders could not be persuaded to undertake the journey again and Chiltern Celtic Dancer was a lone competitor. He travelled from Harwich, care of Margo Dean, who kindly took him on board on her way through from Ireland. I travelled over later with Tim, Ros and Patrick Harris. Seven year old Patrick carried the Union Jack in the Parade, with Dancer at his heels. Sweden made an even stronger contribution on this occasion, with fifty-four entries. Ger-

[194] O.O.M. 186

many mustered five entries and Holland four. Finney Master[195] represented France, on his way to his new home in Sweden, and Margo Dean kept the flag flying for Ireland. A heavy, relentless rain for forty-eight hours rendered the Festival a strength of endurance over adversity for all involved, organisers, exhibitors, judges and spectators alike. That the entire programme ran to time and to the finish was a remarkable tribute to Bent and his team. Marble was no longer alive to delight us, but his progeny were outstanding. In the two year old filly class, eight of his daughters, like peas in a pod, stood at the head of the line. I was proud that Dancer was Reserve Champion Young Pony to the winner, Oxenholm Mandy. Bardot took the Supreme Championship.

It could be said my two visits to Roskilde were an expensive failure, with no success for Cameo and no sale for Dancer. I prefer to look back on them as a worthwhile experience which I would not have missed. I think we learned that a powerful stud of good ponies can destroy the appetite of others for competition, by the sheer weight of their supremacy. I'm not certain whether the 1981 downpour, or the early death of Marble, or some new interest, extinguished Bent's fire of enthusiasm for the Connemara pony. His stud was gradually dispersed, to the benefit of many European breeders, and Denmark continues to breed good ponies, under regular supervision from the Irish inspectors. I found a small morsel of personal compensation in the 1993 Connemara Chronicle in which the Danish report included a photograph of Tegelstrups Baroness. In 1977 the Larsen family admired Cameo and asked whether I might find them a filly of similar quality. As a result Bill Woodley's, Red Baroness,[196] travelled to Denmark where she became a successful brood mare. Tegelstrups Baroness is one of her progeny.

Before moving on to my next journey, a further word about International Festivals. In 1979, Sweden followed the Danish example and staged a similar Festival at Vetlanda. I was not able to be there and, in fact, Sweden is one European country I have not visited. Reports on Vetlanda suggest it was a colourful, successful occasion, but without the aid of a catalogue I am not certain whether visiting countries other than Denmark and Germany were represented. It must have been dispiriting for the Swedes to see all the major awards taken to Denmark by the Oxenholm stud.

Since 1981, two attempts to 'Go International' have been made by the English and Dutch societies. The former did not materialise, for there were no pony entries. However, overseas breeders were welcomed at an evening reception. More of the Dutch 1992 International show later.

It seems to me the cost of travelling ponies overseas will always prove a deterrent to the vision of an International Pony Festival. It is unlikely someone of Bent Nielsons calibre will surface and pick up the challenge in the near, or indeed the

[195] O.O.M. 224-226 [196] O.O.M. 199

far future. It is surely more practical for breeders to meet, without their ponies, at Clifden when they can.

In 1978 the De Rooys invited me to Holland. We corresponded for nearly ten years but had never met. I travelled by sea from Sheerness to Vlissingen where I caught a train to Arnhem. Lily and Marolyn were standing on the platform holding a Chronicle, as a method of identification. Their invitation was planned to coincide with the Nederlands tenth anniversary show which was held at Uddel and fell, by a happy coincidence, on my forty-ninth birthday.

On my arrival at Brummen I was soon at home and became involved with the busy preparations for the show. Time was found to take me for a ride on Chiltern Arctic Venture,[197] which was a happy reunion for me after ten years. I well remembered the agonies I suffered on saying goodbye to her at Yarmouth Docks, not knowing what her future might hold. I could not have wished for a better life for her than she has had with the De Rooys.

Dr. Lockefeer was Chairman of the Dutch Society in 1978. We had met previously over breakfast in a Knightsbridge Hotel, prior to a COPA working party in London. I was made most welcome at Uddel and was happy to renew acquaintance with the Verschurens and to meet other Dutch breeders. I was given a copy of the Society's first stud book and spent the day at the ringside using it and my catalogue to identify the ponies. One hundred and sixty of them were entered. A Jury of four undertook the judging and I enjoyed being a spectator in the sunshine. I rejoiced in Marolyn's success when she won the Championship with her mare Vanessa, bred from April, a Rebel Wind mare whom they had imported some years earlier.

As I look through the catalogue fifteen years later, three familiar names catch my attention. Strike Action, six years old, out of Power Station by Killyreagh Kim, who was 5th in her class; Clifden Charm, out of Clifden Molly by Doon Paddy (my note beside her entry reads, small, lacks presence) Ocean Sky out of Muffy by Errigal Prince (my note, nice pony). She was placed 3rd in her class which was won by Floor Colleen, with Kingstown Swallow in second place.

I was delighted to return to Holland in 1986 to judge at their show, held at Schaijk. I flew from Birmingham to Schipol, Amsterdam, and shared the task of judging with Dr. Eikdenboom. We got along very happily together, our only disagreement occurring when I suggested we leave the indoor arena once it had stopped raining. The Doctor was anxious not to get his catalogue wet if the rain reappeared. I was dismayed at his use of the catalogue which, he told me, was standard practice when judging in Holland, and necessary for breeding information! We were asked to make the awards in Premiums 1A down to E and 2 and 3 in a similar fashion. 3D had to be a very poor pony! Floor Collen was busy in the

[197] O.O.M. 201-202

jumping ring on this occasion, but I discovered three of her daughters were entered. A powerful brood mare, Atlantic Gael[198] by Ben Lettery, had to give way to Van de Arenbosch Frederique by Kimble, in a battle for the Championship.

I was staying with the De Rooys, and so their ponies were not shown. On days either side of the show, they took me to the Flevohof Agricultural Centre, and on a boat trip, from the majestic Rotterdam Harbour, to view the famous Dutch dams at work. I was fascinated by various aspects of Dutch rural life; the general use of bicycles for transport and for pleasure; the many attractive waterways or dykes which carve their way through the flat, Dutch landscape; the beautiful windmills, mostly no longer in use; the small farmland enclosures for grazing cattle and growing maize, all immaculately tidy and well tended.

Imagine my pleasure on receiving an invitation to judge at the Society's twenty-fifth anniversary show at Heteren in 1992. It was billed as an International show and was run under the able chairmanship of Mr. De Jong. The Germans entered a team of eight ponies, with whom they achieved considerable success. Frau Schwartz, together with the Milleders and the Hillnhutters were supportive spectators, and Anne Hemphill and I shared an evening meal with them, only to find we were celebrating Ediths birthday. Daniel and Nicole Blontroch travelled their mare and foal from Belgium and won a ridden class with the mare. Between them they provided the show with an International flavour.

Lady Hemphill and I were joined by Merv Koelewign from Holland for the two day judging stint. Anne and I were slightly ill at ease with the system of judging, each pony coming forward for an individual assessment and using the ice skating mark system. All was totted up on a computer!! We found our Dutch co-judge placed great importance on action, while we were looking for bone and type. It may have been disappointing for Dutch breeders to see twenty-eight year old Ballydonagh Deirdre gain the Championship with the best young pony coming from Germany, Bright Pearl by Flagrant.

Two or three moments were special for me and remain fresh in my memory: the stallion, Carna Gold,[199] in his eighteenth year, resplendent in a set of white bandages, giving a gallant display of Dressage, ridden in a double bridle in which he was clearly ill at ease; April, in her twenty-eighth year and dignified in old age, leading her family of three daughters for judging in the progeny class, Vanessa, aged fifteen, and two cross bred mares aged sixteen and six years; Floor Colleen, a quality lightly built dun mare, whom I had seen competing at each of the three Dutch shows I attended, now in her twenty-fifth year, with a foal at foot and placed third in her class; Atlantic Curragh,[200] in his twenty-fifth year, looking spare but alert and as active as ever, winning the senior stallion class in which he has rarely, if ever, been defeated since he arrived in Holland. And finally, Snoopy

[198] O.O.M. 215 [199] O.O.M. 208-209 [200] O.O.M. 207-208

v.d.Leliard, a smallish flea bitten grey mare who quite took my heart. She won her class and five of her progeny were on the show ground, four of whom won their class.

A special word of congratulation to Mr. V. D. Meijden who entered sixteen ponies and who won the supreme with Ballydonagh Deidre. He paraded his gelding Mancho Van Graaf Janshof, returned from Stockholm, where he was the European Junior Event Champion. Mancho's dam, Clonkeehan Kittiwake, bred Mr. V. D. Meijden a string of good ponies. Kittiwake, I find, is by Strongbow out of Clonkeehan Fuschia whose dam, Nipper of Calla, was out of Calla Lass by Tiger Gill. What splendid old fashioned blood she brought to Holland, and how proud Frances Lee Norman would have been at Mancho's success.

It occurs to me that when Atlantic Curragh comes to the end of his useful life - and that surely cannot be too far off - the time is ripe for the introduction of a new male line to Holland, one with greater bone and substance maybe. Curragh has been a marvellous servant to Dutch breeders, who must be congratulated on the long, healthy and active lives their ponies lead. It would be hard to find a tougher little nut than Floor Colleen by Atlantic Storm.

In July 1978 Mary and Rogers Rutter from the U.S.A. visited the U.K. in search of Connemara youngstock for their Tower Hill stud. They chose the yearlings Cocum Coppelia and Chiltern Copa, and made arrangements for Pedens to fly them to the U.S.A. When Rogers discovered he had to pay the air-fare for a groom to travel with them he rang offering me 'the ticket', together with the opportunity to spend some time with them in the States. They were planning to travel to Kentucky for the World Three Day Event Championships and suggested I might like to go along with them. What a golden opportunity!

I drove Copa to Sussex and the following morning John Williams drove the two ponies and myself to Gatwick airport. The ponies were travelling with a consignment of thoroughbreds and the loading, in their pallet stalls aboard a fork-lift, was trouble-free. As I climbed the steep stairway to the rear of the plane, I was greeted with the wry comment (from the all male travelling groom personnel) "We were expecting Vera". Apparently they had been told a Miss Lynn would be travelling with them and they expected Vera Lynn, no less, or so they said! They tossed me a couple of blankets saying it would be very cold and spent most of the flight playing cards and drinking beer. I spent my time keeping the ponies company the other end of the plane. When we landed at Kennedy Airport, we were greeted by a blistering heat of ninety degrees. The ponies had to go into quarantine and were not allowed to put a toe on the tarmac at the airport. They walked directly onto the ramp of the horse box which was waiting for them. After seeing the ponies disappear through the quarantine gates, we drove to the Rutters home at Sanbornton, returning for the ponies three days later, when they were released from quarantine. I could tell Mary was taking her Connemara breeding very seriously. She had a barn full of sons and daughters of their much loved sire, Whitewood Galway Bay.[201] I was concerned to find their grazing facilities were very limited but their move to Vermont from New Hampshire has given their stud a new look, as indeed have Copa and Coppelia and their progeny.

The following day Mary, the two Nesbit girls and myself, set off for Kentucky in a borrowed 'camper', leaving Rogers in charge of the ponies. The Horse Park facility, which had been prepared especially for the World Championships, was deeply impressive. I could scarcely believe I was treading the famous blue grass turf of Kentucky land. It was good to be there and to wave the Union Jack in support of the British team, though sadly the excessive heat and humidity impaired the overall success of the event. I had a two-fold personal interest in the performance of our team, for Richard Meade on Bleak Hills was the son of our E.C.P.S. Chairman John Meade. And Christopher Collins on Smokey VI was an old friend, who gained much of his early riding experience at my Buckinghamshire yard, helping to break young horses and school point to pointers.

The American Connemara Pony Society was holding its A.G.M. (always a movable feast) at the time of the Event, not far from the Horse Park. We four arrived at the meeting tired and dishevelled after several nights in the camper. It was my first opportunity to meet American breeders, and subsequently some of them visited us at Redgates.

My return flight was from Boston, and in order to catch the plane we had to drive through the night. I remember Caroline (Nesbit) taking over from Mary in the early hours of the morning, sitting behind the wheel in her night dress. I sat beside her to keep her awake. We talked ponies the whole night through and have been friends ever since.

It was Caroline, in her capacity as Chairperson of Region 1, who invited me to the States in 1989. I was asked to present a paper on "Judging a Connemara Pony", and to conduct a clinic the following day. Thirty ponies of every age, colour, sex and height paraded for my critique. It was a generous response by local American breeders, some of whom were aware their ponies might not receive my approval. I found it a very constructive way of sharing my views on the breed and hope that it was equally rewarding for spectators. All this was done within the context of the Society's A.G.M. held at Hamilton, Massachusettes, where I was very touched to be awarded the Crestwood Trophy in appreciation of what I had shared with them. Breeders travelled from far and wide to be there and I treasure a folder signed with a personal note from them all.

I am sure the U.S.A. gives most of us the feeling of affluence and rolling acres. It was a revelation for me to discover how many of those breeding Con-

[201] O.O.M. 114

nemara ponies were doing so with both limited means and facilities. I admired them and what they were achieving all the more for that.

My 1989 visit to America remains full of happy memories; the three days I spent with Caroline, Bob and Chiltern Colm at Wintervale Farm in New Hampshire; visiting old friends Pagey Elliot, Lousie Clough and Sally Green; staying with Cary Kennedy at her lovely Michigan home and sitting on the box seat of her coach behind her impeccable four-in-hand Connemara driving team; flying to Ohio to spend time with Greg Nalepka and his mother, Claire. (Greg spent six weeks at Redgates as a result of our parking the camper in his parents driveway in 1978); renewing friendship with Mary More Gaines of the Finchampstead stud in England, but now living in San Francisco; the car drive from New Hampshire to Vermont to visit the Rutters on a sunlit September Sunday as the autumn fall took off; a trail ride on Greg's Quarter Horse through woodland glades, apparently designed for the pleasure of horse riders; meeting the ponies at Joanie Websters Wild Horse Valley Road, Stonybrook stud, standing contentedly in the unaccustomed dry Californian heat; clambering aboard a ranch horse, complete with Western saddle, to ride behind George Gaines high in the Californian hills, and finally, viewing the beauties of Virginia with Catherine Mack, aboard the comfortable back of Far Aboves Katherine[202] or behind Callowfeenish Thunder between the shafts. I travelled far and saw much, but realise I have touched only the tip of the iceberg. New York State, Georgia, Florida, Missourri, Texas all beckon, with Connemara breeders I would love to visit.

The great distances make one very aware of the difficulties the Society have ahead of them to forge closer ties with the parent body, as members of the ICCPS. Across the country inspections seem an unreal dream, even if they were acceptable to everyone. Would not an inspection of stallions, starting with the next generation of two year old colts be a more realistic objective?

It was Copa and Coppelia who first took me to America and so it is over to them for the last word. In a recent copy of the American Journal I saw a photograph of Tower Hill's Wizard of Oz. The caption said he was by Copa out of Coppelia. In the event I discovered his dam is Springledge Dublin Lass. His owner Kathy Williams, tells me 'Ozie does it all' from leading winning Hunt Teams four miles across country, to carrying Samantha side saddle over fences at A shows, jousting, show-jumping and trail-riding. Ozie has recently been joined at the Williams Funny Farm by Tower Hill's Leprechaun, who is indeed, a son of Copa and Coppelia. I'm sure he will prove as good a performer as Ozie, who remarkably does all his work in a surgical shoe protecting a broken coffin bone.

My very first overseas judging assignment was in France in 1979. Phyllis Meade was invited, and when unable to make the journey, she proposed me as her

[202] O.O.M. 118 + 119

replacement. It was the second show run by Anne Morgan, on behalf of the French Association, at her home Oursieres. It was, she told me, two hundred miles south of Paris in the Loire valley and would be best reached by road. Thus Jo and I decided to make the journey by car, and see something of the French countryside and Paris en route. How grateful I was to have Jo with me when I discovered my co-judge, the Directeur of the Haras du Blois, spoke no English. Jo, a fluent French speaker, came into the ring to act as interpreter.

Oursieres is an elegant French Chateaux, set in its own picturesque grounds and provided a perfect back-cloth for the two day event. We enjoyed the Morgan hospitality and the constant flow of wine from breakfast through to dinner. Unfortunately, I had eaten a sausage roll in Paris (while Jo went up the Eiffel Tower) which did not agree with me, and I carried out my duties with some difficulty.

The in-hand classes were thin and, on the whole, disappointing. We were happy with our champion, the mare Boogie Woogie,[203] a daughter of Bridge Boy, and grandaughter of Cregg Lassie. We placed Idenoir[204] second in the yearling colt class and what a good sire he has proved for French breeders. I remember being more impressed by the riding ponies than the breeding stock, and an attractive cross-country course had been designed in the Chateaux grounds, part of which was woodland. An outstanding performer, Glendunn, by Lambay Fireball, gave an exhibition of how to jump big fences.

On Saturday evening the Morgans welcomed all the villagers, together with those involved with the show, to join them at a 'Mechoi' in their walled garden. Two whole sheep were roasted on the spit, wine flowed more generously than ever, everyone spoke French — and I was still on a starvation diet!!

By the time I returned to France in 1990, the Association show was firmly established at Tours, and on this occasion Steph Brooks was my partner in crime. We met at Heathrow and flew to Paris together. Steph bravely undertook to get us to the station from the airport via the underground. We made the three hour journey to Tours by train.

The show ground facilities at Tours can be compared with those at Stoneleigh in England and I was pleased to find a vast improvement in the quality of pony entered in 1990. A veritable pot-pourri of colourful memories remain fresh from the Tours visit.

Our day began when we met Francois Birkigt and her stallion Cliff,[205] sharing breakfast at the do-it-yourself breakfast bar, with Cliff in no more than a head-collar. And later that day the same owners thirty-three year old Dun Lorenzo mare, Early Sunrise, provided one of the highlights of the show as she trotted briskly round the ring, to show us how well she was and how happy to be there.

[203] O.O.M. 248 [204] O.O.M. 236 [205] O.O.M. 242

The show organiser, the Baronne de Sevein, was on the move throughout the two days looking after all our needs, with her faithful attendant at her heels, her little brown dog, Brusille. Among the competitors I noticed Monsier Corbinus, who showed his handsome sons of Island Earl[206] with great expertise and success. Hubert Laurent was busy with a string of ponies from his Melody stud, all immaculately produced and showing the same characteristics. White Granite, yet another of Marble's charming daughters, was second in her class, with the winning colt foal at foot, and three more of her sons all taking major honours. Monsieur Rene Delaune was showing his pony, Venise Melody, for the first time and anxiously learning the art as he went. His efforts were rewarded with the special prize, donated by the Cocum stud, for the best French bred mare or filly. Rene's photograph, with his mare, was posted to me as a souvenir of Tours. The stallion class generated great excitement as Monsieur Phillibert's stallion, Garryhinch Millrace, bought at Clifden a month earlier, entered the ring with a flourish. He had arrived in France that very day and an exhibitor objected to his presence, in the belief that his papers could not be in order. One minute we were judging him. The next we were not. A major incident was averted when Monsieur Philibert politely withdrew him.

An evening meal with exhibitors brought our busy two days to an end. On Sunday Steph and I were free, and took a taxi to Tours where I attended Mass in the Cathedral. We spent the remainder of a happy day as tourists.

Two visits to Germany came comparatively close together, in September 1987 and in May 1990. This gave an extra flavour to my judging in 1990. One of the few compensations a judge can experience, is the rare occasion when their confidence in a youngster is vindicated, if they have the opportunity to judge the same pony again at maturity. This was my pleasure in Germany.

I found the two shows different in character, although many of the exhibitors were the same. The 1987 show was devoted to Connemara ponies and held at Kargen in East Bavaria. The Hauser family were our hosts and their lovely home, set in beautiful surroundings, provided every facility for man and beast. I drove from Munich to Kargen with the Milleders, who were travelling their two mares in a trailer. It was a sunny autumn day, but by the following morning it was grey and wet, and four inches of rain fell within the next forty-eight hours. The conditions at Kargen compared with those experienced ten years earlier at Roskilde.

I shared the judging with Marie-Luise Schwartz, a charming German lady in her seventieth year, a respected breeder and judge of ponies. We began our task at 8.00 am and had worked our way through twenty-one in-hand classes by 2.00 pm. By this time we were very wet.

[206] O.O.M. 240

The stallion Westside Frank[207] had recently joined the Hauser stud from Ireland. Jurgen Kwade travelled his Clifden Champion, Coosheen Nutmeg,[208] his resident sire, Tulira Nimble Dick[209] and their progeny the long journey from Nordhorn. But it was Herr Dorfler's German bred mare, Marena,[210] who won the Championship. The afternoon session, which did not concern us, was full of activity, with show jumping, dressage, driving and vaulting classes. The winner of the driving class took Frau Schwartz and myself for a drive through the forest and the lovely dried flowers presented to us got as wet as we did. The organisers stuck to their guns and the programme came to a close with an evening celebration of music, beer and hot dogs, held in the dry of the Hauser indoor arena. Earlier Marie-Luise and I returned to the hotel in search of a bath and dry clothes. I have never been able to solve the complexities of bathroom showers and was no luckier on this occasion. The cold water ran vigorously but nothing I did encouraged the hot!

An out-door Mass was planned for Sunday morning with a Choir of male singers in full Bavarian costume blowing their curled hunting horns to accompany the celebration. The rain denied us the out-doors and the indoor arena was a welcome but poor substitute.

The Schonbach Ponyfest is run for the benefit of nine pony breeds. Each Society takes it in turn to organise the event and in 1990 it was the responsibility of the Connemara Society. Lady Maria Levinge was invited as my co-judge. We met in Dublin and flew to Frankfurt together. That evening we shared dinner with the Hillnhutter family who were celebrating the silver wedding anniversary of their parents. We were unexpectedly joined by some almost unknown relatives from the East and found ourselves experiencing a small piece of German history, for it was shortly after the Berlin wall came down.

The following day we were impressed by the busy preparations for the show and the number of ponies they were taking, to include the four required for Kurt's Quarndom driving team. The carriage and harness involved careful stowage, and camping and cooking equipment were needed for the three nights they would be away from home. On arriving at the show-ground each family was allocated an area of grass, for their ponies and their camp. Ponies were fenced off by electric tape and grazed happily in their own groups. The weather was kind in 1990 and it seemed an admirable way of celebrating a Ponyfest.

Maria and I were very taken with Claudia Mattke's champion stallion Carraway[211] and his full sister, Jessica, winner of the barren mare class. The sire Golden Dan[212] had two good winners, his four year old son Glaskoph Golden Merlin and an eight year old daughter Scarlett.[213] The mare, Fortuna, won the progeny group

[207] O.O.M. 174
[208] O.O.M. 166
[209] O.O.M. 165-166
[210] O.O.M. 178
[211] O.O.M. 171
[212] O.O.M. 167-168
[213] O.O.M. 180

class with two full sisters by Bengt and a third by Gonzales. Marena was Champion mare and reserve to Caraway for the supreme.

During the course of my two visits to Germany I detected a close relationship between the German girls and their ponies. They respect the free spirit of the Connemara pony and trust him implicitly. This trust is returned in full and the bond between them is reflected in their relaxed enjoyment of each other. A special, and rather different, relationship exists between Kurt Hillnhutter and Golden Dan but 'trust' is, again, the bottom line.

Herr Impings pair of Dutch bred Connemara mares drawing his coach gave Maria and I a royal drive round the main arena after a presentation of the traditional bouquet of dried flowers. Maria found it hard to believe that Customs would not allow us to take them home.

Later the same year Claudia Mattke and her boyfriend were touring England and arrived at Combe with their tent, primus stove and an album of photographs. A welcome mark of friendship.

My German experience did not begin and end with ponies, thanks to the generosity of the Milleders. The life and works of Blessed Teresia Benedicta, the Jewish Carmelite nun, who died in Aushwitz, had for some time been a source of inspiration and a guiding light as I sought to adapt to life at Presteigne Carmel (she may be easier recognised as the scholar, Dr Edith Stein). In 1987 I was able to follow a few footsteps of her life, by visiting Cologne Carmel, spending some time at Dachau Carmel and travelling to the Benedictine Monastery at Beuron, where Edith had often shared the Easter Liturgy. It was a spiritual pilgrimage which left a lasting impression and one I shall always be grateful to have made.

Memories of my three weeks in Australia in 1980 are of long journeys through drought ridden land, of getting on and off planes at Melbourne Airport, of meeting Australian friends in their own homes and seeing how they bred and used their Connemara ponies. I was driven through Victoria, across Queensland and over the border to New South Wales. I saw a little of South Australia, a lot of Tasmania and nothing of Alice Springs or Western Australia.

I was invited as Overseas Visitor to the Society's Native Pony Show at Bowral in May. Before this event, a busy itinerary was arranged, which included a visit to Adelaide to judge the Connemara classes at the Expo. Only a handful of ponies came forward but they included an old friend (Aylesland) Silver Velvet.[214] An attractive sire, Connemara Park Wilfrid by Clonkeehan Archer, was my Champion. An evening gathering of South Australian breeders was enjoyed by all.

On my arrival at Melbourne I was met by 'M' Kelly and taken to the Storey's home at Arthurs Creek. I found that ponies with the Blandings prefix were scattered, on ajistment,* at Pakenham, on Kate's farm at Neerim Junction, with

[214] O.O.M. 149-150

*Rented Grazing

friends at Thornton or at Eagles Hill. I did not do a head count, but remember a strong-boned full brother and sister, Connemara Park Richard and Judy, and the Blandings born sire by Carna Bobby. As days passed, I discovered it was not unusual for Australian breeders to have a large number of mares and two or more stallions. I marvelled at Kate Storey's ability to keep her finger on the pulse of her pony breeding with so many miles between them. I soon learnt, what I considered to be long drives were a fact of life for the Australians.

After judging at Adelaide I drove with 'M' to her Barwidgee home at Caramut. I began to get the feeling of Australia as we passed through Tailem Bend and Tintintara and called on the Withers at Naracoorte. Barwidgee realised my expectations of a large Australian station, running a herd of cattle - Colin's pedigree Aberdeen Angus - thousands of sheep and a not inconsiderable number of horses and ponies. What a relief to meet the reformed Tulira Colman[215] behind a high voltage electric fence! The last time we met he had his teeth in Island Duke's neck at the Ascot stallion show!!

I returned to Melbourne to fly to Tasmania for my four day visit. I was filled with excitement, rather like a parent about to be reunited with her family. The three Chiltern ponies looked well and happy and the Tasmanian breeders gave me a great welcome. Ann Paterson's Home Vale stud was well ordered, as was my visit. I met every breeder on the Island, saw my first kangaroo, and drove through the 'fine wool' growing country and the lush Huon Valley.

Brisbane and the Clarke Glenormiston home were my next stop and there I was treated to the picture of the Clarke children riding bareback, (wearing crash helmets but no shoes) up and down the steep hillside behind their home, among the avocado pear trees. We had a day out at the Marburg show which gave me a peep into rural agricultural life, where the rhythm and strength of the wood cutters and the massive drought master cattle, with humps like camels, were a novel sight.

Jack Conroy's impressive Shelford Downs was planned as my next visit and I was able to share the journey to Bowral with them. The 600 mile two day trek was prepared by Jack down to the last detail and was a journey I'm unlikely to forget. We travelled in convoy with the car following the horse box carrying four ponies. I became more aware of the ravages of the drought and the arid acres with no stock for miles and miles. Lambs were being sold for as little as five pence each in Melbourne market.

On the morning of the Bowral Festival, I begged a lift into town to buy a pair of gum boots. The drought had broken but, as it happened, only tentatively. I judged the in-hand section with Helen Clift and we had the unenviable task of choosing between Millfields Commodore[216] and Domo Cavallo Praize[217] for first place in the stallion class. Each of the seventy ponies catalogued took part in every event and the stallions showed their mettle across country. Some good partbreds

[215] O.O.M. 140-141 [216] O.O.M. 141-142 [217] O.O.M. 138

made an important contribution to the success of the Festival. All ages (human) were catered for, and the Bar-Bar-Aft event - too complicated to explain here - had everyone at top speed. The Australians certainly know how to get the best out of their ponies.

Carol Paine on Barwidgee Fiddler,[218] winning the geriatrics cross-country event, and Sue Clarke riding in every class with a scalded foot, displayed typical Australian grit. The Saturday evening Barbecue was an informal gathering at which I was asked to talk about the Connemara pony in his native habitat. This was followed by a lively question and answer session.

I left Bowral with the McGowans and spent my last two nights in Australia at their home, Yarrandoo. It was a 400 mile drive which we completed in six hours in Paul's Mercedes. This visit was one of the happiest experiences of my Australian tour. The McGowan triplets, three teenage girls, owned a Connemara pony apiece and they took me for a ride round the farm the next day. In the evening, when Marie discovered my visit had been so taken up with ponies that 'culture' had been overlooked, I was given a tour of the house. The 'Trips' were my guide and together we wandered from room to room finding the works of Australian artists, photographers, sculptors and writers. The memory of the homely warmth and goodness of the McGowan family kept me company on the long, lonely flight home.

When I was unable to accept Sir John Marshalls invitation to fly onto New Zealand from Australia in 1980, I thought it unlikely I would get a second chance to visit New Zealand. Shortly before Christmas 1991, Len Thompson, secretary of the Connemara Society in New Zealand, rang to ask if I would like to fly over for their A.G.M. in March. What a welcome second chance!

The distance between breeders may not be as great as in Australia, but the same problem of bringing breeders and their ponies together exists, in particular in the context of the North Island and the South. The A.G.M. moves from North to South each year, and in 1992 was held at Ambury Park near to Auckland. The day after the meeting a 'Ribbon Day' was arranged, and fourteen ponies and their riders gathered together to enjoy the competition. It was good to see Tom McNamara encouraging the young in this way.

In the afternoon they took an hour off to join me in the indoor school, where we divided the ponies into family groups to represent the four sires, Connemara Park Roy,* Connemara Park Michael, Chiltern Sunsilk and Canningvale Irish Dusk. It made an interesting study and I enjoyed sharing my thoughts on, what I perceived, as the good and not so good points in each group. An overall uniform type among the ponies was a compliment to the sires and I hope New Zealanders were encouraged by my comments.

[218] O.O.M. 141

*Pages 157-162 O.O.M.

Lady Marshall travelled up from Wellington for the A.G.M. and her presence as Patron was very welcome. Fourteen years had passed since she, and the late Sir John, visited me in England to see Chiltern Sunsilk. As my travels in New Zealand progressed I became aware of the valuable contribution their Canningvale Stud, and indeed Chiltern Sunsilk, have made to the breed. Lady Marshall and I were sad to discover Sunsilk had been parted from his mares and was living in apparent exile, in Taupo. I believe efforts have since been made for mares to visit him where he is presently located.

The President, Anne Ryan, invited me to join her when inspecting ponies for registration. This gave me the opportunity to study a number of ponies and their pedigrees at close quarters. A drive from North to South (with Anne) took three days, with two overnight stops and visits to pony breeders en route. This journey gave me a wonderful introduction to the Islands and their scenic variation. Anne was a pilot par excellence and kept me in touch with local practises and history as we drove.

I appreciated the New Zealand hospitality 'with a difference' at Waterfall Farm with the Thompsons, and at Scarteen with the Ryans, where I had the use of a guests bungalow which was standing adjacent to the family home. I discovered how welcome your own 'roof' can be at the end of a tiring day.

It was good to visit Taieri Hore at Blenheim and discover her enthusiasm for the Connemara pony has gained a second wind. She is a skilled horse-woman with a young family who are keen riders, and they use their Connemaras in the best possible way. Her family and nursing commitments do not leave her with spare time for the Society, which is their loss. I was pleased to meet the fifth stallion, Barwidgee Errif, who has spent most of his life out in the New Zealand 'sticks'. He lives with the Turners who are now at Timaunga on the South Island, which, for the first time in his life, gives Eriff the opportunity to cover a few pure bred Connemara mares. Tony Turner is a great enthusiast who does his shepherding aboard a Connemara in preference to the widely used tractor-bikes. His teenage daughter, Phoebe, shares her father's love of the ponies.

Len and Colleen's generous sponsorship of my visit to New Zealand and the Society's invitation to join them as an Honorary Life Member ensure that the future of the breed in New Zealand will remain close to my heart. I doubt anywhere else in the world can offer equines a better way of life than New Zealand. It is a horse 'growing' country and the people are noted horse-masters. The Connemara pony, given a strong Society, will surely have a long term future there.

My visit to Belgium in October 1990 was a fleeting one and did not work out quite as planned. I met two enthusiastic breeders at Clifden, and they invited me to join them for their show in the autumn, with the suggestion I judge the Connemara classes. The organisers of the show, the Belgium Warmblood Society, did not accept their representations, preferring to use the same panel of judges year by year. This Society are custodians of the Connemara stud book and organise the October show at Oud-Herverlee each year, for the benefit of the Belgian Riding pony. Four classes are included for the Connemara pony on the same day.

The fact that I was not judging did not in any way detract from the pleasure of my visit to Belgium, where the Van Rijkevorsel and Blontrock families made me more than welcome, as indeed, did the officers of the show. I was invited to share lunch with them and to sign their visitors book, a very beautiful volume reserved for the signatures of important guests. My morale was restored!

It proved an interesting day and I was invited to 'stand in' on the judging and to pass my critique of each pony to Daniel Blontrock. The Connemaras in Belgium are either Irish or Dutch bred and comparatively small in number. The enthusiasm of the few deserves greater support and it is disappointing for them that the largest breeder, Monsieur Rens, is unable to play an active role in the life of the Society.

The show is held at the headquarters of a Farmers Association which offers every facility for the day. Their offices, dining room, and indoor and outdoor manage, and the park-like grounds, which surround the main buildings, all combine to provide the environment for a successful show. The show classes are followed by the inspection of colts, and of stallions who are relicensed annually. The young entires, presented for the first time, are expected to show their jumping ability either on a lunge rein or performing loose up a lane of poles. A rigorous examination for conformation, action and soundness is undertaken, and owners are justifiably pleased when their stallion passes this test.

My time in Belgium was all too short. We drove fast along the flat, busy motorways carrying traffic to and from Brussels. Along the route I noticed many small farms with few stock grazing but often a large building used for intensive pig rearing. I kept a sharp eye open for the native Belgian heavy horse for whom I have always had an affection. He is not unlike the Irish Draught, though he tends to be 'rounder' in appearance and is roan in colour.

I would like to think I might return to Belgium one day and find fifty ponies instead of fifteen at the Oud-Heverlee show ground.

As we touched down at Helsinki airport I felt I was venturing into the unknown, into a land renowned for its long hard winters, a land of forests, lakes and fishermen. I wondered what I might find there on a sunny August day? A two hour journey by road to Kouvala did not reveal a great deal.

It was 1992 and I was the guest of the Finnish Connemara Pony Society, who, for the first time in their history, were responsible for running their National

Pony Show. It was my task to judge the Shetland and New Forest pony classes, as well as the Connemaras. This was a tall order, made even taller by the request to record my assessment of every pony in every class. It was duly translated into Finnish and relayed over the loud-speaker by the commentator. At the same time, three Inspectors from the Finnish Horse Association were marking the ponies independently for their own records, a little unnerving to say the least! The show was held at the Sururso Equitation Centre and the classes were judged in the sand-based outdoor arena, on two pleasant sunny days.

I was glad when I had the Shetlands (hundreds of them) and the Foresters behind me, and was able to move on to the Connemaras, where I found some interesting ponies. The Champion was a charming yearling filly by Millfields Ensign.[219] Her dam, twenty-one year old Hummelbro Ballerina, won the older mare class and was awarded the Leppa Joy Trophy. The winning foal, by Kasta Scaramouche, Downhill Dewdrop, was quite special and took the Best Foal in Show award. In the final championships. Ametist by Kimble[220] was Champion mare.

I spent a further three days in Finland, and thanks to Tuula and her friends, I saw more of this lovely country than I dared to hope. We visited Tuula's graceful old mare, Espe Aletta,[221] and her family who were grazing peacefully in their summer home beside the lake, surrounded by forest through which the ponies could wander, play, and graze at will. From there we drove on to Lappenrata to see Ardnasillagh O'Flaherty[222] who is now a contented old pony in his nineteenth year. To see him rekindled my memories of him as a four year old at Roskilde, where he performed with great vigour and presence in the Dressage classes.

We returned to Kouvala to load two yearling colts who had competed against each other the previous day. They - and I - were on our way to Iittala, where I was to spend the night, and the colts the next few months. It was a long drive and as, finally, we wound our way along a narrow forest track, I wondered where it might lead … suddenly a burst of daylight revealed a fairy tale green oasis, in which was set the farmhouse home of Kristos, Sari and their three small boys. They farm a small acreage in traditional Finnish fashion, milking a herd of pedigree Ayrshire cows, running a flock of sheep and harvesting their own animal feed. Sari does the milking and works her sheepdog 'Netia', with whom she competes in sheepdog trials. Kristos is fully occupied working the land, (in the few months in which it is possible to work it), and in the winter months he built Sari a magnificent barn for the ponies, which they share with the farm machinery.

Sari prepared an all-Finnish, home produced, evening meal for us, wild duck (shot by Kristos), Finnish wine made from wild berries, and cooked beastings laced with cinnamon and berries. We ate from a long table overlooking the lake

[219] O.O.M. 218-219 [220] See Index O.O.M. [221] O.O.M. 259 [222] O.O.M. 187-188

with Sari's mares and foals grazing beside it and the cows beyond, all encircled by the greener than green forest. I slept like a log in one of the boys bunk beds.

Next morning we visited the boys and their grand-parents at their holiday chalet. I learnt that most families own a similar chalet for use as a summer leisure retreat. On our way to Jamsa, Sari encouraged me to savour a bowl of reindeer soup and we watched a Finnish glass blower at work as we did so.

Outi Talvi and her family own a fine equestrian complex which comprises a barn full of ponies (mostly Connemaras), a spacious indoor riding school, a main house with cooking facilities and lounge (cared for by her parents), and a row of small individual chalets for sleeping accommodation. Outi is an accomplished riding instructress and her centre is a riding holiday-home for many children. Breeding Connemara ponies is twinned with this aspect of her business, and is her great interest. She has recently imported a Danish bred colt by Hazy Dawn,[223] and two mares from England. They were still in quarantine when we visited them. What a strange twist of fate for me to meet Haughmond Danielle, one of Chiltern Manikins very few daughters, for the first time on Finnish soil. Her head bore an uncanny resemblance to her sire.

Outi's boy friend drove me to the airport and on the way he shared Finlands history with me. As I boarded the plane I mused how 'five days on', Finland was no longer 'unknown' but a land in which I felt very much at home. I was happy to see how well the Connemara pony has adapted to the alien Northern climate, where he is obviously giving pleasure to so many.

Donal Kenny's enthusiasm for the Basotho Pony Project and the export of two Connemara stallions to Lesotho, as a contribution to the improvement of the breed, fired my imagination. Whenever we met, Donal discussed the progress of the project and even suggested I should visit to see the work for myself. This seemed a remote possibility.

In 1987 the Prioress of Presteigne Carmel asked me to accompany her on a visit to South Africa. I had a god-daughter living in Cape Town and knew I would be welcome there, while Sister Anne visited Wynberg Carmel, the purpose of our journey. Suddenly the realisation that Lesotho was easily accessible from Cape Town occurred to me. I wrote to Donal and by a remarkable coincidence his annual visit to the Kingdom was planned for the period I was free. I could hardly believe my good fortune.

Donal met me at Blomfontein airport on November 29th and we drove to the capital, Maseru, being allowed to cross the frontier only after a prolonged examination of our passports. Maseru is a strange town, an unusual blend of Africa and Europe. Women were sitting on the pavement weaving and selling their conical

[223] O.O.M. 189-190

hats and rising up behind them, garish neon signs advertising Spar and Gilbey's Gin!!

It seemed hardly possible that at the beginning of this unique journey I did not feel very great. I wondered if I was suffering from too much sun, after a long drive in intense heat, touring the Cape the day before. I did my best to shrug it off as we prepared for the five hour drive to Thaba Tseka. Tommy Ryan, an exuberant Irishman and manager of the project, was our pilot. After an hour we stopped for lunch at the Molimo Nthuse Lodge Hotel which was built on stilts in the side of a mountain, a circular building with the traditional style conical roof. As we left, Tommy remarked 'now we batten down the hatches' and we set off along the dirt track road which twisted and curved like a snake rising from 5,000 ft to 12,000 ft. We drove through the God Help us Pass, and the Robbers Pass, as fast as Tommy could encourage his Toyota to travel. By this time I had a fearsome pain in the back of my head which blurred my appreciation of the beauty of the mountains as we scaled them. The gaping ravines and sheer drops either side of the track, had me holding my breath on every curve. We arrived at Thaba Tseka in one piece where we were greeted by John, one of two young expatriate Irishmen working on the project. His house girl, Tattoo, prepared us a meal and I spent the night in the spare room of John's small house.

I felt a little better next morning and we set off on a tour of the stud in the 'baki' (a four wheel drive pick up) with Gilbert, the stud groom and a native of Lesotho, as our guide. A modern stable block with all the usual offices was surrounded by grassland paddocks, in which were grazing small groups of ponies, divided by age and by sex. There were twelve yearling fillies in one enclosure and I was told the following year they would be assessed and culled, with only the best retained for breeding purposes. They were by four different sires, one of whom was the Connemara, Crockaun Connemara,[224] always referred to as "Croc". We found him grazing with a small herd of mares and foals, on a hillside, looking for all the world as if he was in his native homeland. A 'minder' was sitting motionless in the corner of the field, under his conical hat and colourful blanket, with pencil and paper. At the end of the day his findings were recorded on the large 'covering' chart on the wall of the tack room. Nothing is left to chance at the Thaba Teseka stud. Beyond the grazing paddocks the land is farmed and I welcomed the opportunity of watching the people at work. The men were building a dam, by hand, stone by stone; oxen were drawing a plough; an experiment at growing banana grass was being watched anxiously; women were harvesting peas by hand with a sickle; radish tops were saved in drums and converted to silage; women worked in rows pulling the weed, senecio, (ragwort) from the crop it was damaging. Each task was being done with a smile and I realised I was seeing the best side of life in

[224] O.O.M.266-267

the Third World, thanks to the input of this very worthwhile Irish project. A whole new village has developed around the stud where the people have an opportunity to work.

I learned of the benefits of the Basotho ponies triple gait and how racing the ponies was a great sport for the people. Mare camps were being set up in outlying villages, so that the breeding programme was not restricted to Thaba Tseka. I saw the native widow bird and picked some of their beautiful flora. I walked through a field of lucerne out of which we disturbed a flock of storks, searching for snakes, Tello, our young guide told us, with some relish!

I was not looking forward to the downhill return journey. We did it in double quick time and when we reached Tommy's house at Molimo Nthuse I was fit only for long iced drinks and bed. I was shivering violently every so often and felt rough.

The following day I was taken to the marketing and trekking centre, another aspect of the project, where Milford Hurricane[225] (now a gelding) was at work. A three hour trek had been arranged for me, Michael was my guide and we did no more than walk as my Basotho pony picked his way surely over the rough terrain on a loose rein. We stopped for a short rest beside one of the many magnificent waterfalls.

Donal and Tommy were giving a party for the staff at the Hotel that night. I finally had to admit defeat when my shakes were so bad I could not hold a glass. Next day Donal took me to the Roma Mission Hospital where I sat in a queue among the people waiting for my turn. I saw a Canadian lady Doctor and instant tests revealed a bad kidney infection. Antibiotics and forty-eight hours in bed were ordered. I had twenty-four in an hotel bedroom at Blomfontein, before having to catch the plane home.

For obvious reasons my Lesotho experience was bitter-sweet, but left me with a colourful collage of African memories and Irish kindness, never to be forgotten.

[225] O.O.M.266-267

Tattoo (far right) her husband (seated) and their family outside their small house at Thaba Tseka.

CHAPTER SIX

WHY ISLANDS?

Islands and Connemara ponies have long had a close association. Inchagoill, Lambay, Lough Mask and other smaller Islands off the Western Seaboard, and Bardsey, off the North Wales, Lleyn Peninsular, have provided the nursery grounds for many good Connemara ponies. They have also held a strange fascination for me since my earliest years. The prospect of crossing the sea, leaving civilisation behind me and being self-sufficient presented a challenge to my childhood imagination. In more recent years Islands have held a spiritual significance for me, as places of prayer, of peace and quiet and closeness to nature and to God. I have discovered how all Islands have a similar history of vikings and of celts, of monasticism, hermits, saints and pilgrimages, of piracy and of farming the land and fishing the sea. Their rich and varied past fostered a desire to visit a few of them and to unravel the stories they have to tell.

Bardsey lies two and a half miles off the Lleyn Peninsular and is named after a Viking Chieftain, Bardr. The Welsh title, Ynys Ennli, is a descriptive one, 'The Island in the Tides'. It is these tides which govern the movement between the Island and the mainland. Arthur Strick and Ernest Evans are two men who understand and respect the vagaries of the tides so well.

The history of the Island can be traced to AD 429 when the first holy man, Einion Frenhin is said to have arrived. He was followed by many, many more and Bardsey is often referred to as the Island of twenty thousand saints, said to be buried there. In the middle ages it was an important place of pilgrimage and three pilgrimages to Bardsey were considered to equal one to Rome. In the thirteenth century the Augustinians built an Abbey on Bardsey and the ruins remain to this day. After the Dissolution of the Monasteries it became a pirate stronghold until 1790 when a community of some eighty people was established, men who farmed the land and fished the sea. The lighthouse was built in 1821 and by 1870 sturdier farmhouses were erected and in 1875 a Methodist Chapel also. Religion and secular education were supplied by the minister and a resident school teacher.

In 1972/73 Arthur and Jane Strick took the tenancy of the farmhouse Ty Pellag and the 400 acre farm land, and decided to run a flock of sheep and a herd of Connemara ponies. They became the only 'all the year round' residents on the Island. Lighthouse keepers worked in shifts until automation took over. An Anglican Sister lived a simple hermit life in a farm building until ill health and old age forced her off the Island. Dick Loxton from the Dept. of Zoology, Leeds, spent some months of the year in the Strick's hayloft, and wardens of the Bird Field Observatory live and work there in the spring and summer months.

Life on Bardsey, they tell me, offers a sense of security, a sense of order and a certain freedom. It is a life which can only be lived by those who accept the rule of nature as their master, realising that each and every day's work is subject to the provisions of storm and calm, sun and rain, drought and downpour. To tune into the radio weather station is an early morning priority. The day is planned according to the information it provides. Bardsey has been known to be cut off from the mainland for as long as four weeks, four or five days is not unusual and up to ten days in September can be expected. Everything, a gallon of diesel, a can of baked beans, a bag of sheep ration, has to be brought from the mainland. This is the life Arthur and Jane chose when they took the farm tenancy twenty years ago.

When we see the Bardsey ponies competing at Olympia, at the breed shows and in the Performance Award Scheme with such success, many have little idea of the endeavour it has required to breed and rear these ponies. When foundation stock was purchased, a means of shipping them to the Island had to be devised. Ernest built a sea truck, using an old wreck as its outer skin and driftwood for its body. Before engines were fitted, it was towed behind a launch for the forty minute crossing, carrying two ponies at a time. This is how the first ponies reached Bardsey and how the seventy bred there since, have travelled to the mainland.

I am sure it is beyond most of us to imagine life without a telephone and easy access to a vet. When there were colts for castration two vets on the mainland pencilled Bardsey in their diary for every Tuesday in May and June, in the hope that on one of the eight, a crossing there and back, in a day, would be possible. Sheep shearing and hay making always present their own problems, as does the importance of transporting lambs to the mainland when the market is ripe.

The use of the natural treasures of the Island will soon be a thing of the past. The willow from the withy beds may still be used for making lobster pots but creels[226] (baskets as in Connemara) are no longer required for bringing in the seaweed from the sea shore. Before 1940 each small farm had its 'furziary' where gorse was cultivated, the sweet branches of which were chopped for feed and for bedding, while the woody stalks were used for firewood. The salvaging of driftwood remains a daily routine and anyone staying on the Island is made aware of the collectors right to his collection. This is assembled on the headland, and marked with an initial.

Bardsey both lured me and teased me for seven years. Between 1973 and 1975 I made three journeys to Aberdaran to deliver ponies for the foundation of the

[226] S.I.M. 13

Loose Ends

Dalesman (far L) with John Killeen at Clifden in 1973

Doris Davison's three month old American bred colt Sky Lyne (L)

Ann O'Malley (below) at Ballinafad in 1992

Pat Mulkerrin (Tom) and his wife at home in 1991 (below L)

James Carr and Dimple at Aille in 1992 (bottom R)

C. N. De Courcy-Parry, Kilmeena, Westport, Co. Mayo.

29. Dec. 45

Oh Dear me.

And your "Chronicle" no doubt in the Press and myself filled with idleness and good intent commenced upon a contribution for you and then tore it up and came to Mallow to look after a mansion whence the owner had gone Englanding to grandmother over the festivities. CHELTENHAM.' You guessed correctly. We often wonder if there is any where else in England excepting Cheltenham. Anyway her house was full of books, every sort of books, bed side lights, larders and under-the-stairs filled with luxuries, a daily worker and central boiling. God save us, Paradise enow - after my sordid hut on the coast of Mayo and, well I wrote nothing, until to-day when I submit this to you to do exactly as you wish, no Pride nor Prejudice, just a sorry its no better.

Here, there is, as ever, utmost confusion to such as myself as to what IS a CONNEMARA ? Myself being still in the era of the greys and the duns, nor being able to reconcile to the blacks and the browns. What harm the blue eye... Wales was full of such and the blue eyed dog sees further than his brown eyed comrade.

A Happy and prosperous New Year.. All goes well. Don't be believing the pessimists. Birds sing. Daffofils are out. Spring is upon us in this country of No Winter..

I wish I was looking at your so grim Black Mountains.

God Love You.

Bay de Courcy Dalesman

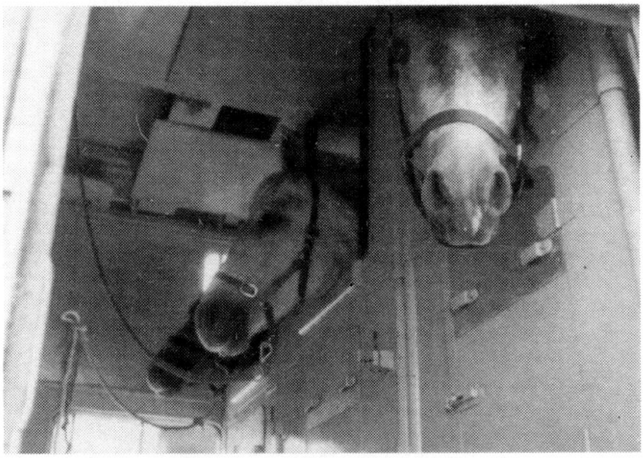

DENMARK – ROSKILDE

The ponies parade at Roskilde, 1977 (top)

Cecilie Williams puts the finishing touches to Fort Viola with Jean Fidler in the saddle (above R)

Chiltern Cameo and Tiercel Witch Hazel in the horse box on board boat (above L)

Bent Nielsen keeps the show going in 1981 (L)

HOLLAND

SHAIJK 1986

HETEREN 1992

Champion pony 1986 v. de Arenbosch Frederique (top L)

Dr Eikdenboom and the author dressed for the weather (top R)

Snoopy v. de Leliard (centre L)

Carna Gold (centre R)

Mancho Van Graaf Janshof (L) European Cross Country Champion, 1992, displays his jumping ability

Travelog

U.S.A.

Tower Hill's Dareena (top L) a daughter of Copa and Copelia, Vermont 1989 and (R) Copa and Copelia are fork lifted into the body of the plane, Gatwick 1978.

At the American Breeders Dinner (L) Hamilton 1989 (from L to R) Catherine Mack, Aladdin Stud, the author, Pat Lightbody (Editor American Connemara) Laura Carpenter, Tre Awain Stud and President.

Tower Hill's Wizard of Oz and Samantha Williams leading the Old Dominian Hunt Team to Victory, 1991 (Below)

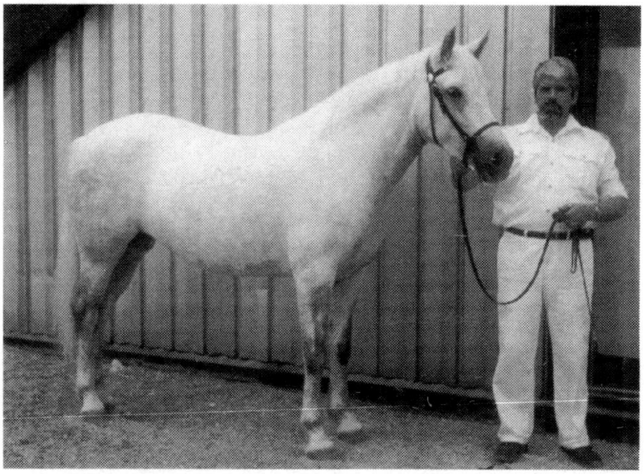

FRANCE

Richard and Anne Morgan (above R) watch the presentation of awards on the forecourt of their home, Oursieres, 1979

Hubert Laurent and his Champion mare Nougatine Melody with Stephanie Brooks at Tours 1990 (above L)

Rene Delaune (L) with his mare Venise Melody.

A six in hand driving team of blue eyed cream Connemara ponies parade at Tours (below)

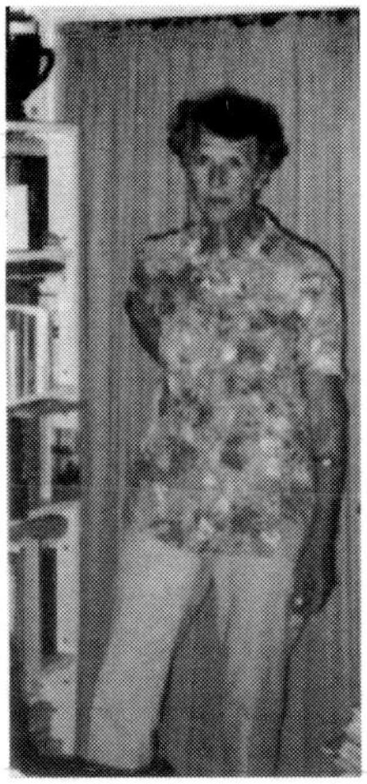

GERMANY

Kurt Hillnhutter at Kargen 1987 (L)

Marie-Luise Schwartz at home (far L)

Seventeen year old Aughnish Dolly and her daughter, Gina, arrive at Kargen after a six week 750 mile ride from Essen (centre R)

A happy gathering outside the Connemara Secretariat at Schonbach 1990 (centre L)

Herr Imping takes the Judges for a drive in his coach (below)

Travelog

AUSTRALIA

'M' Kelly (top) with one of her good part bred performers.

In gum boots under a gum tree!! (above)

Chiltern Saffron one of the Shelford Downs successful team (centre R)

Two full sisters from the Waterfall Stud, Enniskillen and Moycullen (below L)

Tom MacNamara leads Lady Marshall across the ring at Ambury Park to present the rosettes (below R)

NEW ZEALAND

Travelog

BELGIUM

Tulira Smasher attends Oud - Heverlee in 1990 for his licence to be renewed

FINLAND

Sari and Tuula in the new barn at Mayra (L)

Outi Talvi's three year old stallion Brunskaer Mars (below L)

Champion Foal at Kouvala 1992, Downhill Dewdrop (below)

LESOTHO

BARDSEY

Basotho ponies ridden by husband and wife (top L)

Michael, the author's guide, rides ahead of her into the Lesotho mountains (below L)

Arthur and Jane Strick and their assistant, aboard the sea truck with two ponies. (top R)

Bardsey Flora competes in a working hunter pony class. (below R)

Why Islands ?

Mathew and Stephen Tierney prepare to leave Inchagoill Island (top L)

Stephen steers the boat across Lough Corrib. (R)

Two of the Dun Aengus crew steady the rowing boat below, with Inishbofin in the distance. (L)

Anne Marie Burke, Librarian on Inishbofin. beside her Renault (below L)

Heather and Desmond Duane in the 1930's. (below R)

Why Islands ?

Paddy O'Halloran at the wheel of the Dun Aengus (top R)

Murteen Joyce and the author on the pier at Inishbarra (top L)

Murteen with Atalanta, and her foal by Ardravinna Andy. (L)

Murteen and his dog follow An Cailin Fearnuil Fionn across the Island. (below L)

The Omey Races in 1946. Garnet Irwin on Irish Boy by Adventure nearest the camera (below R)

Bardsey herd. Each time I left the ponies on the mainland and only once, on a clear day, did I catch a glimpse of the Island. Would I ever set foot on Bardsey, I wondered? In August 1980 I set off early in the morning with a plan to meet Arthur at Fishermans Cove, Porth. I was on my way to inspect the two year olds for registration. I sat on the beach all day looking out to sea and waiting anxiously for Arthur's arrival. He never came. The tides decreed otherwise. At six o'clock I relinquished my vigil and found B&B in Aberdaran, determined to return to the cove the following morning in the hope conditions might have improved. My excitement as a boat appeared on the horizon was great. Arthur handed me some oil skins and, before I clambered aboard, told me, tersely, there was no guarantee he could return me to the mainland that day. After a seven year wait and Bardsey within hailing distance, I was unlikely to be deterred by such a threat, and threw all discretion to the wind of the Bardsey Sound. I sat on the wooden box at the rear of Arthur's fibreglass dory, powered by an outboard motor, and hung on for dear life as we hit one big wave after another. It was an exhilarating experience as the spray covered me and the sun glistened on the waves.

I did return that evening, as planned, but on two subsequent visits I stayed overnight and savoured the stillness and beauty of the Island. On our evening walk we collected mushrooms and driftwood, we wandered among the Abbey ruins and the tombstones and I pondered how life might have been on Bardsey in Celtic times. We went to bed to the tune of the weird cackling of the Manx sheer-waters.

In August 1994 Arthur and Jane will relinquish their farm tenancy and move to their retirement home Brynawelon, Aberdaran. They will see Bardsey's mountain top from the large sitting room window on a clear day and I guess they will leave, at least a small part, of their heart on the Island. Happily, Jane hopes to continue breeding ponies and I'm sure the Bardsey prefix will remain a popular one. These ponies have brought their own special ingredients of natural hardiness and good sense to the mainland, which has been widely appreciated. The stallion Bardsey Zodiac[227] and the mare Bardsey Flora are particular favourites of mine. They have proved themselves in so many different ways.

Today the Island is the property of the Bardsey Island Trust whose aim is to preserve the natural history, religious sites and all the Island's amenities for the benefit of the general public. Some of the old houses have been refurbished and are let in the summer months to those interested in bird watching, making a group or a private retreat, or just getting away from it all.

> *"This is truly an enchanted place, a place of ancient blessing - when words have blown away like spindrift you hear in the deep hearts core. On a night when the*

[227] O.O.M. 49

Manx sheerwaters come homing in to their burrows, with that weird sound like the cry of souls in torment, or with the rim of the moon on the mountain's shoulder and an opal-eyed Connemara mare etched against a creaming sea, the present tense with all its imperatives seem curiously irrelevant".

(Christopher Wordsworth - Island Britain)

Inchagoill is one of the largest of the many Islands which dot the twenty-seven mile stretch of the magnificent Lough Corrib. This Island extends to three hundred acres and is no longer used for grazing but is managed by the state as a forestry. The stallion, Inchagoill Laddie,[228] was the one to lead me to this particular Island, for it was here that he and his companions spent their earliest years. Three sires carrying the Inchagoill prefix were registered in 1936, Laddie, Star[229] and Hero.[230] It seems likely they were three from a bunch of ten colt foals who were placed on the Island as weanlings for eighteen months or more. The stud book tells me they were inspected at Oughterard in 1936 and were passed for registration as being 'of true pony type and fair average quality'. I believe they were inspected in the autumn and not used until three years old. Laddie was the only one of the three to realise the Inspector's confidence. I made the pilgrimage to Inchagoill to follow his childhood footsteps, to see where he had grown and matured. The men who handled this stallion inspired me with their respect and admiration for him.

While I was living at Errislannan in 1981 Jim and Nora Walsh arranged for me to meet Stephen Tierney, who lives at Portacarron on the shore of Lough Corrib. He would, they said, be happy to take me to the Island. I left Errislannan early, in bright sunshine, on October 31st, Halloween, and I feared a stiff breeze might deter Stephen from making the trip. I had been warned that the gentle and inviting Corrib can become as tempestuous as the Irish sea, in a short space of time. I looked anxiously at Stephen's face as he considered the extent of the breeze before offering me a set of oil skins, while he and his son Matthew prepared the boat for the trip. He suggested I might find the journey too rough. I stoutly defended my ability to cope, in view of having survived the crossing to Bardsey! The twenty-five minute journey was smoother than that to Bardsey but we were well doused in spray as the rowing boat, powered by a Yamaha engine, forged its way across the Lough. Matthew made the boat fast at the pier, on the East side of the Island, and I stepped ashore full of wonder at the magic, this place, "The Island of the Devout Stranger", held for me.

I visited the ruins of the two Churches, one built in the fifth century and dedicated to St. Patrick, the other in Romanesque style, built in the twelfth century, and known as the Church of the Saint. Its doorway is of special archeological

[228] S.I.M. 151 & 152 [229] S.I.M. 164 [230] S.I.M. 151

interest, as is the stone pillar inscribed LIA LUGNEADEN AMMACC LIMENEUH, The Stone of Lugnaedon, son of Limenuen, who is believed to have been the nephew of, and navigator for, St. Patrick himself.

As I wandered among the trees I mused on the days of the warring O'Connors and O'Flaherties who had, without doubt, used the Island as a refuge or a vantage point upon occasion. I thought of the many who had lived and prayed there; of the Archbishop of Tuam, Maurice O Nioc, who died on the Island in 1128; of those who had been shipped from Cornamona for burial in famine times and whose headstones were mostly in the form of a cross with the inscriptions worn away; of the Guinness family who owned the Island as part of the Ashford Estate from 1855 onwards and of Sir Benjamin's 'Coffee House' built for use on his visits to the Island; of Thomas Kinneavy who lived here in the late 1800s through to 1906, where he carried out his trade of boat building; of Tom Nevin, a caretaker for the Guinness family, and the last man to live on the Island, until 1946. I fancy Tom must have been the ponies' guardian between 1934 - 1936. I hope he enjoyed their company and occasionally threw them a bale of hay. On the West side of the Island lies a delightful short stretch of golden strand and I wondered if the ponies occasionally took a roll here followed by a refreshing drink from the Lough maybe? Not for the first time I wished Bartley O'Sullivan was still alive to tell me how they travelled the ponies to and fro, whether they all survived their tough childhood and oh so much more.

I left the Island with some sadness. No longer was one likely to meet a hermit at his prayers or a herd of ponies grazing peacefully, nor yet the Guinness family on their way to a coffee party. No, it is now an Island of trees, with an empty thatched cottage and a graveyard of the tombstones of forgotten souls. As we drew away from the pier I acknowledged the beauty of the trees in their autumn splendour of bronze and gold, which shone against the more recently planted evergreens. Time passes and things change and I was pleased to have my sadness relieved by a happy celebration around the Tierney table. For our lunch mother, Mary, prepared boiled ham, complete with jacket potatoes and turnips straight from the garden. Young Stephen Tierney had baked a succulent Halloween Brack from which he retrieved a ring and his brother a matchstick, symbols of marriage and of poverty. I am never likely to forget Halloween 1981.

My introduction to the Island of Inishbofin was made in Thomas McHugh's[231] homestead on the Truska beach. His haunting recital of 'The Cleggan Disaster of 1927' in which nine lives from Inishbofin were lost, remains a vivid memory. Some four years after the publication of Shrouded in Mist I read a small notice in the Connacht Tribune. Inishbofin was planning to open a Library, and gifts of books were invited. I posted a complimentary copy of Shrouded in Mist

[231] S.I.M. 69 & 70

directly to the Island and some months later received a thank you from the Librarian in Galway, my old friend Mr. Sharkey*. One day, I told myself, I will make a pilgrimage to Inishbofin, this time to meet the Librarian and to see my book on her shelves. The opportunity to do so did not present itself until 1991.

It was an overcast May morning when I booked my return ticket to Inishbofin at the Cleggan Spar grocery store. Beside the pier I discovered the Dun Aengus, no longer at work on the route to the Aran Islands, but the property of skipper Paddy O'Halloran who provides a reliable daily service between Cleggan and Inishbofin. There was only one other passenger, and together we assisted the crew of two load their cargo of mail, milk, timber and various groceries all marked 'Days'. The sea was like a mill pond and the clouds rolled back as the sun broke through, and the thirty-five minute journey passed all too quickly. I was rather dismayed to find we had to leave the Dun Aengus 'off shore,' climbing down some steep steps into a wobbly rowing boat below. This conveyed us the final few hundred yards to the Island beach. From there I walked directly to Days Hotel. As I munched a 'fresh from the sea' crab sandwich, I meditated, from the long sun-lounge window, on the fishing boats rocking gently on their moorings in the harbour ahead. When I enquired where I might find the Librarian, I was told Ann Marie Burke would very likely be in the grocery store, and that her yellow and blue left hand drive Renault would be outside if she were. I received a great welcome and was taken directly to her library-come-craft shop, which is housed in an Island cottage no longer fit for habitation. Part of the roof had blown off in a recent gale and there was chaos within, for which Ann Marie apologised. Not a sign of books, never mind Shrouded in Mist! She was waiting for an overdue delivery of shelving and the books had been temporarily moved to a safe place. As we sat and talked, I admired Ann Marie's determination to provide the Island with a book service and visitors with a choice of crochet and knitting work from home-spun wool, worked by herself and her family. I gave her a copy of Out of the Mist before she took me to meet Eric and Joanne Elliott. This is how my friendship with Inishbofin began.

I had two hours to spare before the return journey and decided to walk to the East beach, taking the single stony track, one and a half miles long and half the length of the Island. The Island appeared deserted, apart from the cow with her calf, the tethered tiddling lamb, the donkey sharing a small enclosure with his turf cart, and the blackbird who was busy building a nest in Mr. Green's house, recently destroyed by fire. I eventually reached the Monastery ruins which are surrounded by a large untended graveyard overrun with nettles, thistles and yellow iris. The gravestones are a motley selection of the old and the new, the plain and

* Mr. Sharkey assisted with the research for Shrouded in Mist

the ornate. I felt an urge to spend three months on the Island and give it the care it once had.

Following a disagreement at the Whitby Synod in 663 Saint Colm took his Monks from Lindisfarne to Inishbofin. They built a simple wooden Abbey which was replaced in the fourteenth century by one of stone, with walls three feet thick. It was built by the O'Malleys - who had seized the Island from the O'Flaherties - but it suffered the same fate as all others at the time of the Dissolution. Bofin Islanders were the last to submit to English forces after a long seige, and they have often displayed similar courage, when disaster has overtaken them.

In 1907 the Island was taken in care by the Congested Districts Board, who apportioned out the land, and the people have lived from farming and fishing ever since. They have their own new Church overlooking the harbour (completed in 1914) and have a resident priest. There is a post office and two hotels, Days and Doonmore, together with a Youth Hostel and farmhouse B&B. The population today averages two hundred. I was reluctant to leave the Island after so happy and peaceful a day. I sat on the quayside watching a small boy flying his kite and listening to the clang of the crane engaged on the building of a new pier. My eyes rested on the Leenane Head, the last of the Zulu boats recently restored and lying in the harbour. As Paddy O'Halloran steered us home aboard the Dun Aengus I was filled with wonder at the Connemara mountains standing against the skyline, a mauve hue with their tops shrouded in a heat mist, which bode well for the weather next day. I was determined this would not be my last visit to Bofin.

I kept in touch with life and events on the Island through the Inishbofin Inquirer, produced by the Daisy Press on the Island, care of Eric and Joanne Elliott. In 1993 the Islanders decided to hold an Arts Festival from May 14th - 23rd. It tied in very nicely with the Connemara Pony inspections and I arranged to spend the first two nights of the Festival on the Island. A marquee was erected on the harbour's edge and works of Art were exhibited in the two Hotels. I enjoyed a poetry reading by Richard Murphy, the launch of 'Inishbofin through Time and Tide' and listening to a very moving reading from "An Evil Cradling" by the author Brian Keenan. A variety of music was provided and played the whole night through. I travelled to and from the Island with the wife and son of the respected Belfast artist T. P. Flanagan and their friend Maria McAlister. I met Ann Parker (nee Bodkin, a member of the twelve tribes of Galway and proud of it) and sat beside her on the quay as she whittled away at a carving of a Penal Cross. I stayed in a B&B on the East beach with Helena Schofield and begged a lift in her old banger to the harbour. We hurtled along the single stony track, she, ... with a one hours driving lesson on the beach as her licence, and no silencer on her exhaust!! The new pier had been completed and arrival and departure made easy.

Their most distinguished visitor, President Mary Robinson, however, arrived by helicopter! Long may Inishbofin survive the thriving community it is today.

Readers may wonder if Bofin has any relationship with the Connemara pony at all and I have to confess it is but a slim one. In 1894 the Congested Districts Board sent the Welsh pony stallion, Movement*, to the Island. He spent only one year there before being moved to Clare Island and he was not replaced. This suggests breeding ponies was not one of the Islanders interests. Horses were used for farm work and for drawing the currachs and fish catch from one end of the Island to the other. And swimming races on horse or pony back were a one time popular sport.

Matt Duane of Rhus, Moyard was mentioned in Shrouded in Mist as having an interest in the Connemara flaps. Among Garnet's treasury of photographs I found one of Desmond Duane on a grey mare called Heather,[232] born in 1922 and registered number 158. She won first prize at Oughterard show in 1930 but was better known as a winner of the swimming races between the mainland and Bofin. In the recently published book "Inishbofin through Time and Tide" an article from the Connacht Tribune on August 14th 1937 is included with a set of photographs recording a new type of horse racing which had become known 'as riders through the sea'. The horses raced from the mainland across to the Island, the course being a mile (surely considerably more than this?) of the Atlantic. An interesting postscript follows "In the 1930s a unique sport was the horse race to Lyon - not an event for the faint hearted - the horse with rider aboard had to swim from the strand of Duac through the tide to Lyon. It is remembered that a horse belonging to Martin O'Halloran of Middlequarter, was winner, on at least one occasion. This ceased after a few years."

Iomaidh, St. Feichan's seat, is the small Island of Omey, which is separated from Claddagduff on the mainland, by a tidal stream. In days of yore when the water receded the strand beneath provided one of the most popular, and possibly the best, of the Connemara race tracks.[233] The going may have been 'dead' but the surface was even, more than could be said of Clifden, Oughterard and elsewhere.

One of my tasks, while working with the Carmelite sisters, was to welcome their guests and to help them feel at home in the Monastery atmosphere. This was not always easy, because my interests seldom coincided with those of the guests. Imagine my surprise and pleasure on meeting Bea Dufort who, she told me, was married in the Catholic Church in Clifden and had spent her honeymoon in Connemara. She had happy memories of a day at the Omey races and remembered her surprise at the sight of one girl jockey riding alongside the men. Could I tell her who it was she wondered? (It could only have been Garnet!).

* S.I.M. 37 [232] S.I.M. 82 [233] S.I.M. Ch. 5

Talking with Bea rekindled the mental pictures I had drawn of these original race meetings, when writing their story for Shrouded in Mist. Later she sent me a set of photographs taken at Omey in 1946, which portrayed various stages of the day's play. They reminded me of the informality of the riders dress, the wide variety of the contestants, and the side shows which complemented the racing. I have often wondered from whom Michael O'Malley borrowed the racing silks for the posed photo of Cannon Ball[234] in his 1912 booklet?

To visit Omey, you walk or drive across the strand following a line of markers which indicate the route you can safely take, once the tide is out. As you make your way, directly ahead of you lies the long stretch of the Omey graveyard, with lines of Celtic Crosses reaching for the sky. In olden times it was the custom to bury the men on one side and the women on the other and it is here the victims of the Cleggan disaster were taken for burial in 1927.

This small island, two and a half by two miles, boasts two lakes and a Holy Well, valuable fishing grounds and shelter for fishing boats from all the prevailing winds. On my visit I met Martin Davin who farms a sixth of the Island but who does not reside there. I found him repairing a dry stone wall and he told me the one hundred and eight 'loose' Island acres were poor in growth and minerals. He was content to make hay in the traditional cocks and zealously guarded a pair of corncrakes in one of his enclosures. Most of the homesteads on Omey are now holiday homes with no more than half a dozen resident families. It is a tranquil place with the Islands of Bofin and Shark lying to the North of it and Turk and Turbot to the South. The strand no longer hosts race meetings or regattas, but the happy memory of them lingers on – for just a few.

My final Island pilgrimage was to Inishbarra, Bearchan's Island, said to be named after a saint of this name. It lies between Lettermore and Gorumna. When the tide is out it can be reached on foot, by Bothar na Noilean, the road of the Islands, which is a series of very rough causeways and fords. This is the way Murteen Joyce walks his ponies to the mainland when it is time for them to visit the horse, attend the inspections or be taken to the fair.

In 1990 I met Murteen at the Derrynea inspection point with his three year old golden dun filly by Ormond Oliver. He named her An Cailin Fearnil Fion, a brown haired girl who walks like a man. Murteen told me he lived alone on an Island, with his ponies and his dogs for company, I promised myself I would visit him one day. The following year I drove out to Lettermore with Padraig in the hope of finding him on the mainland. We had no luck. His mare, Atalanta, by Atlantic Storm was hobbled and grazing with her foal in the garden of his mainland bungalow. We visited Murteen's neighbours, the bar, the store and the pier

[234] S.I.M. 76 + 77

but no one had seen him that day. He was, surely, on his Island and we had no means of reaching him there.

The following year a prior arrangement was made for us to meet Murteen on the mainland. We found Atalanta and her latest foal sharing the garden with four brand-new upturned currachs,* crafted by Murteen, one of the few men who continues to practise this skill. Together with his brothers, Murteen is renowned locally for the many currach races he won in his youth.

In order to reach the jetty where his rowing boat was moored, we had to negotiate a slippery rocky surface which, for me, was no easy task. It was a damp, misty Sunday afternoon as Murteen rowed us towards Inishbarra and I experienced a strong sensation of living in another age. As we drew alongside the unfinished pier, which towered above us, I wondered anxiously how we reached dry land. It was only thanks to the strong hands of Murteen and Padraig that I did so.

We found An Cailin Fearnil Fion grazing peacefully among the cattle. She was within a few days of her first foaling. Murteen loosed his dogs and they accompanied us on an Island walk. It was hard to realise Inishbarra had supported one hundred and forty seven inhabitants as recently as the nineteen thirties. A school house was built in 1934 for thirty-five pupils and remains a solid building in its squareness. We passed the ruins of the old grocery store and the occasional derelict, empty homestead. I learned how the many stone walls - no longer kept in repair - marked the individual holdings, which enabled each family to be self-sufficient, growing potatoes and oats, grazing the cow and making the hay. It was an Island which had once been full of life, free of worldly riches but rich in its own natural resources and beauty. Today the grazing rights are shared by one or two, and Murteen lives in the only habitable house. In the winter evenings he carves miniature currachs by his kerosene lamp and by day he continues to make oars for the fishermen who continue to use them. The day will come when Murteen's strong arms will no longer be able to row to and from the Island and he will live in his mainland bungalow.

My visit to Inishbarra fulfilled all my childhood dreams of Islands, no engine on the rowing boat and no civilisation on the Island.

Several Islands remain on my visiting list; St. MacDara's, the Island which is treated with much respect by Connemara boatmen and whose early Christian Oratory was restored as recently as 1975; Clare Island, closely associated with the colourful Piratess, Granuaile O'Malley; Birkmore where Carna Dun[235] spent at least a year of his young life, safely out of harms way; tiny Horse Island where the heroic race pony Battleship[236] had the occasional holiday; and lastly Lambay which I have yet to visit.

*A currach is a Connemara fishing boat built in timber covered with canvas and tarred.

[235] S.I.M. 189

[236] S.I.M. 80

Inchagoill, Lambay and the Lough Mask Islands have not seen a pony hoof-print for many a year. Soon Bardsey, and no doubt, Inishbarra, will join the roll call. I am sad that Connemara ponies will no longer roam the Islands which have served them so well. Island life has challenges and hardships, which few are ready to accept in the twentieth century. It is a life which something, deep within me, has always hankered after.

Dan Llanigan O'Keefe judging a working hunter pony class in the U.S.A. 1970.

CHAPTER SEVEN

HERE AND NOW

Certain things never change. The Twelve Bens remain an immovable monument, the Atlantic Ocean continues to sweep the Western Seaboard, the traditional turf cutting, fishing, seaweed harvesting and pony breeding remain an important part of the Connemara culture. But the fabric of life in the region has not escaped the wind of change, which has brought mechanisation, tourism and the benefits of membership of the Common Market in its slipstream. The thatched white-washed cottage has been replaced by the modern bungalow, dotted here, there and everywhere. The donkey has almost disappeared, bicycles are used less and less, cars and tractors are common forms of transport.

Clifden, described by Kathleen Villiers-Tuthill in 1860 "*as a town who stands with little hope in the future, prepared to struggle quietly on, hoping to keep body and soul together*" is now a thriving tourist centre, for at least four months of the year. The first Connemara guide I bought (from 'Moon's of Galway' for 21p) was printed in 1960. Clifden, was said to have five hotels, with one pending, four registered guest houses and four supplementary ones. If you walk the streets of Clifden today and explore the surrounding pockets of civilisation, it is clear to see an explosion has taken place.

I wonder how many of us remember the awkward bridge on the approach to Clifden, now consigned to a lay-by; or the rough field running alongside the Owenglen River where the 'Flaps' of old were run, and which is now the Clifden Glen estate of modern housing; or the building of the telephone exchange and more recently the mart, which portray the march of progress so powerfully?

An even more significant dent in the rural life of Connemara is the recent decision of the Tuam Archdiocese to redeploy their priests. I read how Father John Garvey, a leading figure in the life of the community of Cashel, has left not to be replaced. This is a sad sign of the times. Where would the Connemara pony be today without the leadership and support of Fathers White and Cunningham? Their role as Parish Priest far exceeded the spiritual welfare of their flock. This was clearly demonstrated in their support of the Connemara Pony Breeders Society. Canon Moran, respected Vice President of the Society for the past forty years, is likely to be their final link with the clergy.

It is against this background of change that the Connemara pony and its Society have had to adapt. How easy a task has it been? The day came when the pony was no longer an indispensable part of the peoples' way of life, either as a means of transport or to work on the land. As he was relieved of his heavy work load his physique improved. The breeding policies of the Society slowly filtered through and a better class of pony emerged. A new 'use' of the improved pony and a market for him had to be found if the mares were to be retained and used for breeding.

In Volume VI (1947) of the Connemara Pony Breeders stud book, mention is made of the first ponies exported to England as riding ponies. The following Volume (1951) welcomed the first consignment of ponies leaving for the U.S.A. It is apparent from the report in Volume VIII (1955) these exports were insufficient and that the Society remained seriously concerned about the future of the breed. A paragraph from the 1955 report reads as follows:-

"Mechanisation, even on a small scale has penetrated even to Connemara. The pony, in common with the working horse, is beginning an uphill battle for existence and in recent seasons there has been a noticeable drop in the number of Connemara pony foals being reared ...... apart from the few who ride and jump ponies the younger generation show little interest in these favourites of earlier years. Can something be done to counteract this trend?"

By Volume IX (1963) a more cheerful outlook is reflected in the following extract from their report: *"The many letters received by the Society from all parts of the world demonstrate the general interest invoked in the breed. While no business follows from most of these enquiries, useful connections are formed and sales of ponies often result. In addition to exports to the U.S.A. and Great Britain a few ponies have been exported to Portugal, France, Jersey and the Isle of Man."* It was at this time Stanislaus Lynch, Michael O'Connell and John Daly stepped into the arena. They must given credit for seizing the opportunity the export trade offered them, which, at the same time, provided sales and encouragement for the pony breeder in Connemara. The people and the ponies moved into a new era, the results of which are the ponies we see in Connemara today.

There is little doubt that, today, the ponies lead a softer life. It is rare to see a herd, or a family of ponies grazing far away from home fending for themselves. As the traffic has increased their grazing habits have become more tightly controlled. No longer do the stallions cover many miles beside a bicycle on their stallion route. They travel by trailer and have little opportunity to flex their muscles or to feel the weather on their backs.

I believe, though I may well be a voice crying in the wilderness, that too much regular housing and over feeding, out of season foals born in March and early April, and breeding for extra height, all constitute a threat to the enduring qualities of any native breed of pony. Only history will show how successful this period of blending what is best from the old with the demands of the present, has been.

Those who cherish the Connemara pony rely on the young generation of enthusiastic breeders to be torch-bearers and to find the best way forward for this native pony.

I have just returned from the Clifden show. Where better to assess the health of the breed than there? The usual feast of ponies enthralled me and I was rooted to my concrete seat from 9.30 a.m. until 8.00 p.m. It was a long day and I did not fully digest its contents until later, when I spent two quiet hours with my catalogue and my memories.

It is three years since Mervyn Kingsmill left Connemara, and yet his name continues to dominate the catalogue, and his progeny were in the ribbons throughout the day. John Luskin has a splendid three year old son[237] who was second in his class. Michael Hefferman's four year old son, Streamstown Larry, who was third in the open stallion class, is almost, but not quite, a facsimile of his sire at the same age. Patrick Conneely had fourth in the four year old class with Lehid Colleen. Petie Molloy had his two consistent Kingsmill daughters, Jennifer Rose and Dooneen Castle, in third and fourth position, in an evenly contested class. Perhaps the pick of his prize winning progeny was Henry O'Toole's Castle Dame, winner of the seven to nine year old mare class. There were many others in the front row, if not in the first six.

Abbeyleix Owen was well represented, but his son, Callowfeenish Mairtin, out-stripped his successes, with his three attractive daughters taking a first, second and third place in their respective classes. His winner, Padraig Hynes' yearling filly, out of Village Colleen, (a daughter of Mervyn Kingsmill) went on to win the Youngstock Championship. My eye was taken by at least three useful progeny by Smokey Duncan, all carrying a rich, dark dun coat colour. The new-comer, Village Boy, (by Mervyn Kingsmill) supplied the winner of both the colt and filly foal class which bodes well for the future.

The mares whom we have come to rely upon to provide winners at Clifden, did not let us down. Village Star had children and grand-children taking honours in every other class, and I'm happy to hear she is in foal again this year. Gloves Misty provided the winner of the two and three year old stallion class,[238] the second in the yearling colt class, and had a very charming three year old filly forward, who could do no better than seventh place in her class. Muffy was well represented by two generations. Joe Gorham's eighteen year old mare, Silver Fort, was placed third in her class and provided a daughter[239] to win the five to six year old mare class. Callowfeenish Beauty provided two good grand-daughters, both out of Gillian, who stood third and fourth in their different classes. A six year old grand-daughter of Charm Bracelet,[240] gave Bridget Snow success with a good

[237] Ard Ri Cunga [238] Canal Misty Fionn [239] Kingstown Silver [240] Lakeside Lady

second place. And Frank Joyce* had two of Wireless Wave's daughters in the ribbons.

A popular winner was Noel Sweeney's three year old filly, Irishtown Beauty by Abbeyleix Fionn. Rumour had it she was bought by the Sweeneys because she presented a threat in the show ring to Bay Lass, their home bred filly. Bay Lass is out of Brown Lady and is thus a grand-daughter of their great little mare, Trixie. Bay Lass was second in her class last year, and fourth this.

I am sure I could find many more interesting family connections but must content myself with an unusual one with which to finish. Val Noone's Crusheen Connie, who won the two year old filly class, is out of Nansín Bhán II[241] who is by Rebel Wind out of Majestic Lass and she, Majestic Lass, was bred from a blue eyed cream, Molly Ban, in 1956. Molly Ban's dam, Nansin Ban, was by Camlin Cirrus, a stallion son of Garnet's mare Ciro, who was never greatly favoured as a sire. I returned from Clifden in time to judge the Mountain and Moorland classes at the Monmouth Show where I discovered my winner of the Working Hunter Pony class was a five year old gelding son of Nansín Bhán 2nd and a half brother to Crusheen Connie. What remarkable tricks breeding can play upon one.

The stallion class did not disappoint us, with twenty of the twenty-seven entered in the ring for judging. The judges initial number one, Streamstown Larry, was a surprise choice but in the final assessment he had to give way to Coral Prince, a son of Murphy Rebel,[242] looking fresh and well at the end of a busy season. He had covered seventy-one mares, travelling to most of them, and had won his class at six shows without taking the championship at any one of them. Abbeyleix Fionn, looking equally well, gave him a run for his money before taking second place and Abbeyleix Owen in his twenty-second year stood a gallant fourth. Other interesting entries were Seafield Silver Rogue, apparently a reformed character in new ownership, Island Lad, a son of Kilgreaney Lad, who did not make the front line but looked a workmanlike individual, and Smokey Duncan giving a mannerly display but sadly placed the wrong end of the front line. Perhaps Supreme Champion SIRES should never come back to the Clifden ring?

As the Championship sash was tied around the neck of Coral Prince, in a fading light at 8.00 p.m., my thoughts drifted back to 1966 when Bobby Bolger's father, Jack, had stood in the same Championship spot with the mare Heather Mixture. There was no sash in those days but we may depend the Jan Harald Koechlin Cup was as richly prized and hardly won. I am sorry Jack was not there in 1993 to see Bobby produce and handle the Champion in his own inimitable style.

*Callowfeenish Rose & Cailin Ciuin

[241] Bhán is occasionally spelt Ban or Bawn

[242] See Ch. III

Further impressions gained at the show are a useful guide to the present and to the future. The foal classes received bumper entries, a sure reflection of the generous grants available to breeders during the last three years. Forty fillies and thirty-two colts were catalogued and they were the produce of no less than forty different sires, another point worth reflecting upon. An unplaced filly foal made £1,500 on leaving the ring and the best fillies (at a realistic price) will, I'm sure, find homes easily enough. But what of the colts? I fear many may not make their true value come Maam Cross Fair time. The weakest entry was in the four year old mare class with twelve out of sixteen presented, and this could be a concern in the future. Where oh where have all the good Killyreagh Kim[243] mares disappeared to? Malachy Gorham's Kingstown Molly was the only one in the ribbons and there were no more than four further entries by this good sire in the catalogue. The older mare class was its usual vintage self but, sadly, the weather had deteriorated by then and the gloom set in. What a pity the Dublin Champion, Owens Lady, was not forward, but it was good to find Milford Wren looking as well as ever with another strong foal at foot. It gave me great pleasure to see a nineteen year old daughter of Carrabaun Boy[244] standing in second place.

At the end of the day a spectator remarked "*What has happened to the mares? A stallion has taken the Championship four consecutive years! This is unusual in Connemara.*" I don't know the answer to that one .... Has the quality of the stallions improved I wonder?

While it is good to welcome many new handlers to the Clifden ring it is equally good to see that some of the old timers are still prepared to take the floor. Willie Diamond had ponies in more than one class; John Joyce was there with Heather Cuckoo, a daughter of Wise Cuckoo;[245] Michael Gould showed Brilliant Dawn, a grand-daughter of Brown Alice;[246] Thomas McHugh brought Ballinahinch Gold from Truska, (after an absence from the Clifden ring for some years it was apparent to all how happy he was to be back). And the veteran stallion custodians Josie Conroy and Tommy O'Brien produced their ponies in their customary style. The middle order were well represented by John Luskin, Joe Gorham, Padraig Hynes, Petie Molloy, Jimmy Canavan, Festus King, Henry O'Toole, Eddie Madden, John McLoughlin and many another.

Among the younger generation I noticed Malachy Gorham, Ciaran Curran, Gary Bolger, John Mullen, with his successful filly Fairy Hill Lass, Mary Ward and Stephen Heaune to name but a few. I missed the faces of two regular competitors this year, Sam Morrison from the North of Ireland and Michael King from Derryeighter, Errislannan. I hope they will be back next year. It is the breeders show and it is they who make it the success it is.

[243] See Ch. III [244] O.O.M. Index [245] O.O.M. 284-285 [246] See Ch. III

Clifden has been the home of the Connemara Pony Breeders Society show since 1947, an uninterrupted forty-six years. It keeps to a traditional date, the third Thursday in August. Galway and Oughterard are long established major shows which cater for the Connemara pony, Galway in July and Oughterard at the very end of August. Oughterard hosts a Ridden Championship final for the breed each year and is a popular show.

During the last twelve years there has been a mushrooming of smaller shows in different regions of Connemara. Roundstone was the first, appropriately enough, for it was here the very first Society show was held in 1924. Since then, the Ballyconneely show has become a major fixture in the Connemara calendar and is held in mid July. I hear reports of shows at Claddaghduff, Maam, Spiddal and Carraroe. Shows are an encouragement to breeders to handle their ponies and present them in good condition. And they offer an occasion to gather together for a day out, as did the flaps of old.

A few owners have set a standard of well produced ponies and we have come to expect and appreciate this. Nonetheless, overseas visitors would regret the loss of a certain informality and 'time for all' atmosphere, which pervades the show ground. Too much professionalism and fierce competition would reduce the charm of the occasion.

Claregalway show in 1993 came closely on the heels of Clifden and was held the following Sunday. It was my pleasure to join Anne Hemphill for the judging of the Connemara classes. A good number of ponies came forward, some of whom had been in the ribbons at Clifden. Inevitably the form book was given the odd shake up. It was another day for the ponies, seen by two different pairs of eyes. Entries were made on the day and the trailers rolled in with parking allowed at the ringside. There was bunting to greet you everywhere and in the centre of the ring, tubs of brightly coloured flowers, plus a wooden garden seat, warmly welcomed by foot weary judges. In other areas of the ground, Irish Draughts, Riding ponies, cattle, sheep, goats, donkeys and human babies were being judged, but show jumping was conspicuous by its absence. There was more than enough entertainment without and I do not remember a happier days judging in 1993.

As readers will be well aware the Society stallion list has been a valuable reference throughout my research and I regret its demise. By 1990 the number of Society owned stallions had dwindled to seven and, understandably, a list has not been printed since. It signifies a change of emphasis in the breeding pattern in Connemara today. No longer are breeders totally dependant on the Connemara sire who passes his door on a regular day a week. There are several privately owned sires in the region and most people have some form of transport. I hope and believe there will always be a number of Society stallions available to the one pony man, who remains content with the sire nearest to home.

The Clifden show used to include a Confined stallion class which encouraged the Society stallion custodian to bring his charge, but this has gone the same way as the list. Nonetheless four of the six stallions currently owned (1993) by the Society were shown at Clifden. Abbeyleix Owen, looking as good as we have seen him in spite of advancing years, receives every care with Jimmy Canavan at Moycullen where he continues to cover a good number of mares each season. Smokey Duncan should need no introduction, having been Clifden Champion in 1990 and his stock have already received a mention. Village Boy, bred by Paddy King out of his Abbeyleix Owen mare Village Grey[247] is by Mervyn Kingsmill, and is an interesting newcomer. He is a substantial pony in every respect and carries a valuable selection of genes from which the breed will surely prosper. He is in the care of Tommy O'Brien of Canal Stage and I marvel at the way Tommy negotiates his way out of his yard, towing the trailer with its heavy load with his long wheel base van and immediately taking on a hairpin bend which leads to the main uneven track. Village Boy is always eager to enter the trailer and the journey obviously holds no fears for him. The fourth pony, Moy Johnny's Pride, was brought to Clifden by the ever enthusiastic O'Malley brothers from Carraroe. He is not very big but oozes pony character, has excellent bone and a pleasing personality. He is by The Kid,[248] who left some good ones behind him before going to Italy. His dam, Castle Park, another by Owen, was a particular favourite of Johnny Lee's and it was Michael Lee who ran the colt on to inspection age for the Society.

A second son of Mervyn Kingsmill was cared for by Nicola Musgrave until inspection, but is now at Moycullen with Michael Lee. Matchmakers Lad is different in type to his half brother Village Boy and will need time to mature. Moy Hazy Cove,[249] the last of the six, is in the care of Dudley Nee at Camus and he travels to many of the mares in outlying areas from Screebe to Rosmuc to Carna and back to Cashel. I was disappointed to find only one or two of his progeny were entered for the show. While he may not have many good looks to leave behind him, he too can boast an interesting pedigree and is likely to bequeath his progeny with good hard bone. We must hope that one of these six sires will prove to be a Rebel Wind or a Killyreagh Kim for the year 2000.

The Society's approach to the inspection and registration of colts is taking a new twist. In 1993 a very limited number of colts were passed for registration from an unusually high number who were presented for inspection. It has been decided that, as from 1994, colts will be inspected at three years of age rather than two, as has been the practise hitherto. These two changes in policy will mean a tighter control of the male lines and it remains to be seen whether or not this will be to the advantage of the breed. It may well deplete the gene pool too severely, and rely too heavily on the discernment of the inspectors. While no one would question the

[247] O.O.M. 77 [248] O.O.M. 260 [249] O.O.M. 190

importance of the selection of future sires, to achieve the right balance will, I believe, never be easy. The list of approved sires in Ireland in the 1992/93 season contained the names of one hundred and sixty eight ponies, which indicates the number has risen significantly in the last three years. It may be considered more than necessary and be partly responsible for the latest decisions.

For the last three years it has been mandatory for all stallions to be blood typed and the same procedure is used for other stock when parentage is in doubt. This must be a welcome step forward. In 1992 and 1993 eighteen carefully chosen Connemara stallions were sent to Phillip Scott of Ballina, Co. Mayo for Performance Testing. From all the reports received this has been a successful and useful exercise and the ponies have proved themselves in the way you would expect.

While Performance Testing of stallions is standard practise in some countries on the continent, I wonder how relevant and profitable it is for Connemara sires in their homeland? Does he have to prove himself in this way when his progeny have been doing so, on his behalf, around the world for so many years? I doubt we will learn anything new about his temperament or his natural jumping ability, which he transmits so consistently to his progeny.

The Connemara pony inspections are an event in the Connemara calendar which I would not willingly miss. It is the very best time of year to be in Connemara, as the region slowly comes to life after the long winter. The first foals are arriving and the stallions getting to work, the gorse is in bloom and the grass greening and it is before the avalanche of tourists and coaches arrive. It can be a time of sunshine and of warmth, or sunshine in company with a biting wind. Either way it is good to travel the Connemara roads, to meet the people, and to assess the current crop of two year old ponies.

On my first spring visit which, alas, was not until 1989, the inspections extended to two and a half days, commencing on Tuesday at Barna at 9.00 a.m. and concluding at Tully Cross by midday on Thursday. By then, the region has been well covered and owners have had every opportunity to present their ponies. They meet the inspectors at a cross roads, or at a Bar, or on a village green or up a side road, no matter where, the expectation of meeting old friends or making new ones is a very live one for me.

By 1991 the route had been fine tuned and reduced to two days. A veterinary team, Ned O'Flynn and his wife Liz, are present to sketch mark the ponies identification points and to measure them. The inspection of stallions has been moved from Maam Cross to the Clifden show ground, where a number of fillies also come forward. Moyard, Letterfrack and Tully Cross are no longer meeting places, but ponies from these outlying regions have the opportunity to assemble at the Musgrave's farm at Cleggan. Roundstone is an attractive centre with the ponies lining up with their backs to the harbour wall. The Ballyconneely Catholic Church car-park is always bursting with ponies and trailers, as are Moycullen and Barna. Gortmore has been substituted for Screebe and ponies walk in from Rosmuc to meet the Inspectors there. The long drive out to Kilkerrin and Carna may sometimes seem less than rewarding, with no more than three or four fillies to be seen. I believe they are small nuggets of gold and trust the three we met in 1993 will still be there in 1995.

Memories of the 1993 inspections are a blend of new faces which recall the old, of ponies whose grand-parents are familiar names and of ponies whose obedient trust in those who handle them never ceases to amaze me.

Bal Mahon (short for Bartholomew, a name he did not fancy!) travelled two geldings by Abbeyleix Owen and Coosheen Finn, from Kinvara to Barna. They were tightly packed into a low cattle trailer with their ears pinned to the roof and a peep hole below eye level. It did not appear to worry them a mite, and they walked in and out more than once without demure. Another unique mode of travel was used by the Conneelys of Glentrasna and they, surely, must be descendants of Ed Conneely,[250] stallion man who cared for Charlie, Heather Bell and Derry Boy long ago. The fillies travelled the five miles to Maam in a turf trailer drawn by a tractor. It had no roof, slatted wooden sides and a steep high ramp to negotiate. The wind blew through their manes and they appeared to appreciate their birds eye view of the world as they made their journey, with Michael Conneely at the tractor wheel.

At Derrynea I discovered a filly by Aille Hasty Bits and, knowing he lived in Loughrea, I asked how Grey Aoife, the filly's dam, had met him. The answer was simple. Aille Hasty Bits had travelled through Carraroe with the Leahy trekking train, and had covered Grey Aoife as he passed by. Shades of Cannon Ball on his way home from Athenry* market earlier this century.

One of my favourite meeting points is the Leam turning, the one which Cannon Ball took so often on his weary way home at the end of a long day. Beyond Henry O'Toole's old cottage stands a proud new house built by the Sullivans, which overlooks Cannon Ball's grave marked by a thorn bush. I was encouraged to make the steep climb to the top of the mountain behind the Sullivan's house to meet Leam Little Lady having seen her two year old filly at the Leam turning. The climb was well worthwhile.

Two young handlers, whose faces were new to me, brought their fillies to Leam the same day and how welcome they were. Geraldine Geogeghan is a daughter of Jack and grand-daughter of the one and only Mark. Her neighbour and distant cousin, Martin, is also a Geogeghan, and I notice he showed an eighteen year old Bridge Boy mare into third place at Claregalway show. It seems entirely right that one of the few remaining Bridge Boy mares in Connemara should be breeding for a Geogeghan.

[250] S.I.M. 141 & 167-168 * S.I.M. 147

The greatest surprise the two inspection days provided was the white grey stallion, eight years old, by Thunderbolt and looking for all the world a son of his father. He had spent his eight years as companion to an old man in Mayo, who bought him at Maam Cross as a foal. He had never covered a mare and was over fat, but I hoped the inspectors might consider him. His T.F.C.* (yes he did still have one with Phil MacDermott's signature) told us his dam was Inverin Rose Bud by Ben Lettery. Alas, he did not measure up to the inspector's idea of a sire for 1993.

During the course of the two days, sixty-eight fillies, twenty geldings and one colt were registered. I guess as many or more again were registered at centres outside Connemara. These were held in April in Counties Carlow, Kildare, Meath, Westmeath and Mayo and in Sligo, Claregalway, Northern Ireland and Donegal. Stallions were inspected at three points only, Rathmoylan (Co. Meath), Rockmount (Claregalway) and the Clifden showgrounds.

My visit to the Clifden mart the day after the show, did not persuade me this is a satisfactory way to sell Connemara ponies. However, I accept one breeders comment that a number of ponies were sold who would not otherwise have found new owners. More ponies appeared to pass through the ring unsold than sold and the auctioneers had to work very had indeed to achieve a sale. Old fashioned I may be, but I regret the passing of the Irish Fair which used to be a regular and popular fixture on a certain day of the year in towns up and down the country. So many of the traditional aspects of Irish life are valued and maintained. Why not the Irish Fairs I wonder? I see no mention in the show catalogue of the November Clifden Fair and fear many ponies taken to Maam Cross will again be sold under the hammer.

A new generation of breeders has emerged North of Galway. Tony Donnellan of Gort who advertises twelve brood mares and two stallions at stud; Stephen Joyce of Ballymoe who acquired the stallion, Westside Fred, in the Spring; Pat Finnerty of Clarinbridge who has old Calla Boy running with his young fillies and Eamon Burke who has some nice mares on the Headford road out of Galway and who was lucky enough to have the use of the Clifden Champion, Coral Prince, in the Spring of the year. I was pleased to see another newcomer, Gabriel Murphy of Ballinrobe, at both the Clifden and Claregalway shows. He took third place with Robe Mist at Clifden and Reserve Champion with an older mare at Claregalway. He has the young stallion Thunderball to use on his mares.

Another comparatively new herd of mares has been established at the Letterfrack National Park where an interesting collection of four has been breeding for the last ten years. Two of the mares are by Marble, a third by Carna Bobby and the fourth, Funny Girl, has a rather obscure pedigree. Five fillies have been

* Temporary Foal Certificate

retained, a wise move, as all the mares are twenty years or older. Two 'Park' bred fillies have been sent to the Killarney National Park and all the colts have been sold on. Those at Letterfrack are in the care of Madeline Hannigan.

While new breeders, such as these, suggest a healthy interest in the Connemara pony, it is the small farmer in the heart of Connemara with his single mare who is as essential to the future of the breed as he was to its foundation. Many of them have remained faithful to their mare line and may require encouragement to continue doing so.

The Parish of Carna, whose arms enfold Loughconeera, Kylesalia, Ardmore, Kilkerrin, Moyrus, Mynish, Alnabrone, Callowfeenish and Rusheenamanagh, is now almost a forgotten land and the home of a forgotten people, so cut off is it from the rest of the pony breeding population. It is hard to believe this was once the very centre of things. The show was held at Carna; Joseph Mongan T.D. was Vice President and mine host at his popular Mongans Hotel; the Parish Priest, Father Adams, was a member of the Council and John Walsh was holding a Society stallion at Kylesalia. Today a few valuable old mares and their steadfast owners remain. Two of the mares are by Ben Lettery, one by Dun Aengus and another by Thunderbolt. What precious commodities they are and how important it is that their owners be encouraged to retain a filly from each for the future. I am fully aware how easy it is to accept a tempting offer from someone passing through who has a persuasive manner and a deep pocket.

The Society were aware of this danger as long ago as the nineteen-fifties and I quote from Stud Book Volume VIII *"In 1954 the Society inaugurated a scheme aimed at encouraging breeders to hold their best young fillies to breed from and thus counteract, to some extent, the drain on high class breeding stock due to export. The fillies selected under the scheme cannot be disposed of before they are six years old and the owners must comply with certain other conditions laid down by the Society. A liberal cash subsidy is paid if the conditions are fully complied with."*

It was one of the Ben Lettery mares, Snowdrops Surprise, who provided this year's Clifden Champion. How good it would be to see a stallion in the region again with a *"liberal cash subsidy"* going to the custodian to make it worth his while! A pipe dream maybe, but what an exciting one. My research and writing during the last twenty years has constantly revealed the value of the ponies bred in this region. Inspections, stud books, performance testing and pony shows all rely on the retention of tried and tested mares such as those bred in this nether land of Connemara.

Bartley O'Sullivan wrote in his paper on the Connemara pony in 1939 *"Bred and reared under natural conditions, subsisting for the greater part on the scant rough herbage of mountain and bog, inured to hardships of the most severe order, these ponies are truly in great measure a product of their environment."* This is as relevant today as it was then. That is not to say the pony bred in Bunowen, Cashel, Recess or Oughterard is any less valuable than the one bred in Carna. Each area has its own line of ponies, its own identity, and each one is important to the future prosperity of the breed.

When writing his report earlier this century, Professor Ewart spoke of Connemara as one large stud farm. A stud farm is a provider of stock, and for a native pony, this means stock which is bred and reared on his native heath, the native heath of Connemara being unlike anywhere else in the world.

The recent generous and largely indiscriminate subsidies to mares in a certain age group and to breeders of foals from registered ponies may, I fear, prove to be a double-edged sword. Is there not a grave risk the market will be flooded by too many indifferent youngstock? I hope time will prove my fears unfounded, but surely a more selective distribution of the incentives would have been of greater benefit to the breed? Is it not primarily the small farmer West of Galway and the Corrib who requires help and encouragement to sustain his pony breeding commitment?

1990 saw major changes in the Council Officers of the Connemara Pony Breeders Society. Donal Kenny resigned as President and Eamonn Hannan was elected in his place. Phil MacDermott left her post as Secretary and Marian Turley was appointed in her place. Other familiar faces left the Council to make way for new ones. Much has been achieved in the three years which have followed. The Society's office has been moved to Clifden which is welcomed by everyone. Volumes XVIII (1988) and XIX (1991) of the stud book have been printed. Blood typing has become standard practise. The ICCPS rules, which Donal Kenny compiled and introduced, have been adopted with some refinements. Delegates from overseas Societies meet in Clifden in August every year to report and discuss their problems relating to membership of the ICCPS. The responsibilities of the Connemara Pony Breeders Society, as the parent body, are immense. Eamonn is an enthusiastic and energetic President, Marian a very popular Secretary and Dermot Power a diligent Chairman of Inspectors. As I write, there is a lively trade for the ponies and all those interested in them move towards the year 2000 with confidence.

There are already welcome signs that ponies are moving freely within the Common Market. Germany has been generous in leasing sires to Denmark. Mares have travelled from Ireland to visit an English sire. Finland are looking into the possibility of leasing a sire for two or three years. It is good news that a consignment of ponies, which includes the sire Little Rascal, has recently travelled to Belgium. There will, I'm sure, be many more opportunities for breeders who are prepared to make use of them.

I have a store of golden memories from my visits to Connemara during the last thirty years. On each occasion the same thrill of excitement grips me as I wind my way through Oranmore (I refuse the option of the new by-pass) beyond which I am greeted by a first glimpse of the ever beautiful Galway Bay. The same leap of joy rises in my throat when I see a group of ponies by the roadside, grazing freely as a family, knee deep in boggy sedge with a lough to the right of them and a mountain to the rear of them. To meet Martin Finnerty with his mare Turlough Beauty beside the bicycle, on his way from Rosmuc to Camus to visit the horse or to discover an old mare (with a striking resemblance to Thunderbolt) grazing beneath a wall, heavy in foal, with her ten month old colt still at her heels, are occasions which have given me equal pleasure. I sought the owner of the old mare and her passport told me she was Lily of the Valley. Her dam, Ardmore Small Pearl, foaled in 1972, by Thunderbolt, was a remnant of his first spell in Connemara, two years of which, 1970-71, he spent with Stephen Curran at Ardmore. How remarkable to find a grand-daughter with such a strong resemblance twenty years later. Turlough Beauty and Lily of the Valley are two of the hidden pony population in Connemara today. I hope there are many more of them. I could continue reflecting down memory lane but something tells me it is time to shut the lid of my Pandora's Box.

Writing these Reflections has been a source of joy and, at times, a source of frustration at my own shortcomings as a writer. I hope they have conveyed my appreciation of the West of Ireland, its native pony and the work of the Society. My thirty year association with the Connemara pony in his homeland, around the world, and in my own paddocks, has given me immeasurable pleasure.

On Lambay

ENDPIECE

Sadly - for me - this is the last time I shall travel through the mist with the Connemara pony. I hope there will be someone to pick up the pen and carry on where I have left off in thirty years time. Writing has become a habit and I toy with the idea of a tribute to Arctic Moon, which would double as a stud owners hand book. Friends suggest an autobiography and I am aware as I grow old(er) of the unusual contrasts in my life and of the diverse people who have influenced me. The idea of sharing it is tempting.

As I worked on the final chapter of these Reflections, Chiltern Royalty and Chiltern Quota achieved their remarkable double at the National Pony Society Centenary show, Royalty as Champion Connemara and Quota as Mountain and Moorland Ridden Champion and Olympia qualifier. My memories of buying their dam, Common Market, exactly twenty years ago, remain very fresh. I viewed her from a distance, shrouded in mist, and as I watched her cross the rough, boggy heath, her freedom of movement gave me the incentive, to take a chance, and buy her. Her dam was registered Free Trade in 1962. What better way to close the pages of this book in 1993.

Vanessa Compton on Royalty with Quota beside her.

N.B. Notes on ponies from Colour Plates pages

Ireland (1) Bobby Bolger with his four year old stallion, Coral Prince, Clifden Champion, 1992. (2) Heather Wright and Margo Dean's Ard Conneely, Champion Clifden, 1992. (3) Sean Feeney at Roundstone show in 1987 with his brood mare Lettermacoo Sweetie (Killyreagh Kim - Cool Waterfall). (4) Sian Carson from Carlow, on her versatile pony, Grange Homing Pigeon, (Grange Bobbing Sparrow - Home Lover) on the cross country course at Punchestown in 1992. (5) Abbeyleix Owen in 1991, twenty years old. (6) Killyreagh Kim in 1992 in the last year of his life, twenty-five years old.

Europe. Two of Germany's successful sires; (7) Martin and Patricia Schmidt's fifteen year old Bengt. (8) Kurt Hillnhutters Golden Dan in his twenty fifth year. (9) Karin Johansson from Sweden on her Champion Three Day Event pony Bell Cazo (Brantshammar Bell Dara - Castagne) (photo Henning Vibak). (10) Mr. Van de Meijden's Champion young stallion, Marvel Van Graaf Janshof (Carna Gold - Ballydonagh Deirdre. (11) Palands Bajas the first Connemara stallion born and bred in Norway to receive a licence. Owned by Vermund and Grete Lyngstad, he has a Danish pedigree (Lina Vesterkovs Cecil - Oxenholm Bush Baby). (12) Francois Birkigt and her stallion, Knockadreen Holiday Cliff, receive their prize at the French breeders show at Tours in 1991, from Padraig Hynes. (13) Kasta Scaramouche celebrates Christmas in Finland with a little friend. (14) Frederiksminde Hazy Marvel, a two year old Danish bred colt by the popular sire Hazy Dawn out of Oxenholm Mandy, owned and bred by Gunner Harder Andersen.

England and the U.S.A. Two oustanding **American** bred Performance ponies (15) The mare Balius Kerry Blue (Gilnocky Drumcliff - Polaris Suzanne) owned by the Webster family of the Stonybrook stud (photo Ed Moore). (16) Mr. and Mrs. E. Harris's stallion Hideaways Erin Go Bragh (Hideaways Erin Smithereen - Hideaways Centrefold) (photo Christine Tchir). **In England** (17) Cocum Camelot the English bred, well travelled, son of Thunderbolt and Cocum Prima Donna. Camelot spent two seasons with the Pilgrim's stud in Kent, two on Bardsey Island with the Bardsey herd and is currently (1992/93) in Scotland at Mrs. Simpson's Sydserf stud. (18) Blagdon Haze (Macnamara - Blagdon Dawn) and Amy King, regular winners in Working Hunter pony classes and cross country events. (19) Lewcombe Rondeau (Lewcombe Piccolo - Lambay Sunrise) Champion Mountain and Moorland Working Hunter pony in 1992, owned by Mrs. Leonard and ridden by Melissa Broadfield (photo Pleasure Prints). (20) Shipton Sea Imp (Chiltern Curlew - Shipton Ocean Legend) owned by Mrs. Furness, Ridden Champion ECPS show 1993, taking part in the York and Ainsty Team Event (photo N. B. Jones).

The Antipodes and Africa (21) Gilbert, the stud groom at Thaba Tseka, Lesotho, identifies a filly by Crockaun Connemara. **New Zealand** (22) Canningvale Sunrise by Chiltern Sunsilk ridden by Rosheen McNamara. (23) Anne and Thady Ryan (left and far right) with their mare Glenormiston Dingle Bay (centre) and her two gelding sons by Domo Cavallo Praize and Barwidgee Eriff. (24) Chiltons Way (Chiltern Sunsilk - Ridgeway Gem) a successful competitor in all disciplines, with his owner Elizabeth Mayhead of Taranaki (photo Barbara Thompson). **Australia** (25) eight year old David Moore on six year old Tinderry Clancy (Homevale Quantas - Robinhill Silver Jubilee). (26) Bindt Heard from Victoria on Ballintine Londonderry (Toorigal Danny Boy - Barwidgee Mallow). (27) The twenty-one year old stallion Toorigal Danny Boy (Errislannan Sparkler - Errislannan Columbine) (nearest the bull) running with his mares at Roanoke in a dry season.

Clifden 1993

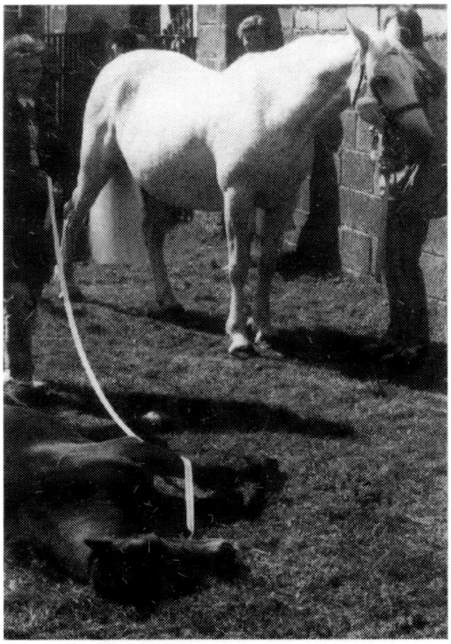

Padraig Hynes' Youngstock Champion, the yearling filly by Callowfeenish Mairtin (top L)

Crusheen Connie winner of the two year old filly class (top R)

A mare and foal at rest, Clifden 1990 (above L)

Festy King runs out his mare, Bunowen Castle Lady, for the judges (above R)

It looks like Padraig Curran and John Brennan to me! - Striking a deal in the traditional way (L)

(Connacht Tribune)

Clifden 1993

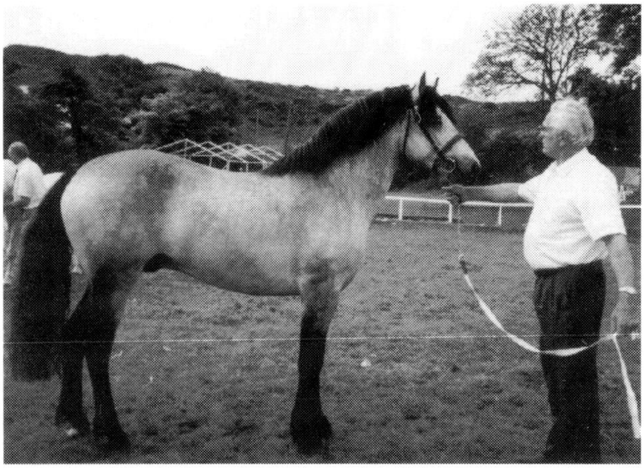

John Luskin and his three year old Ard Ri Cunga, second in the two and three year old colt class (top L)

Sean Dunne and Abbeyleix Fionn, second in the stallion class (top R)

Bobby Bolger's Champion, the four year old stallion Coral Prince (above L)

Malachy Gorham with Kingstown Molly, third in class 14. (above R)

Deborah Laffey with Kildunn Mayflower second in class 14. (L)

(Class 14 is for mares ten years and over. In 1993 Molly was fifteen and Mayflower nineteen years old.)

May Inspections

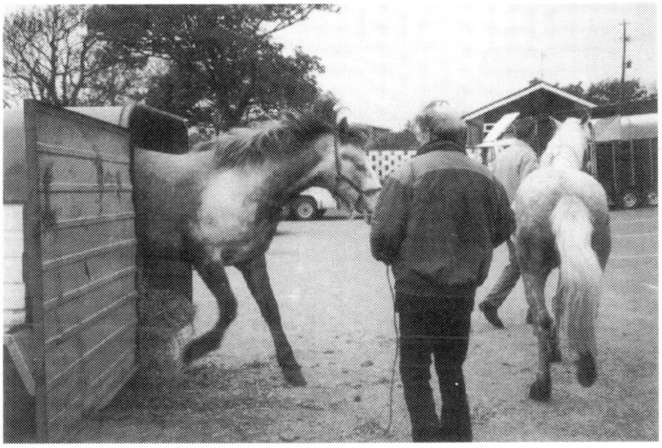

Geraldine Geoghegan and her filly at the Leam inspection point (top R)

Bal Mahon and his two geldings at Barna. (top L)

Liz O'Flynn measures a pony for Jack Geoghegan (Joe Gorham and Dermot Power in the rear) (above L)

At the Derrynea inspection point (L)

The O'Sullivans with their filly at the Leam turning (above)

May Inspections

Eamonn Hannan with his two year old filly, out of his mare Spratt Lady (top L)

Marian Turley collects the registration fee (top R)

Bartley O'Malley with Moy Johnny's Pride (above L)

Tommy O'Brien with Village Boy (above R)

Martin Finnerty and Turloch Beauty (far L)

Martin Mulkerrins looks on (L)

APPENDIX

AUS - Australia
CAN - Canada
FR - France
HOL - Netherlands
NZ - New Zealand
SWE - Sweden
USA - United States of America

BEL - Belgium
DEN - Denmark
GER - Germany
ITY - Italy
SAF - South Africa
SWI - Switzerland

G - Gelding
T.B. - Thoroughbred

I.D. - Irish Draught

Abbeyleix Apollo (FR)
(Carna Bobby - Blue Moon)
Abbeyleix Cypress
(Doon Paddy - Silver Birch of Abbeyleix)
Abbeyleix Delphinium
(Coosheen Finn - Callowfeenish Dolly II)
Abbeyleix Fionn
(Ashfield Sparrow - Fionnuala)
Abbeyleix Molly
(Cilciarain - Clifden Molly)
Abbeyleix Owen
(Kimble - Queen of Diamonds)
Abbeyleix Ri
(Errigal Prince - Queen of Diamonds)
Aille Hasty Bits
(Grange Finn Sparrow - Ballymacrath Beth)
Ametist (SWE/FIN)
(Finney Master - Dame Grainne)
An Cailin Fearnil Fion
(Ormond Oliver - Atalanta)
April (HOL)
(Rebel Wind - Violet Ray)
Arctic Moon
(Carna Dun - Canrower Lass)
Ard Conneely
(Ard Bridge Time - Ard Khana)
Ardan (AUS)
(Inver Rebel - Leam Lassie)
Ardilaun
(Carna Bobby - Ganty Jane)
Ardmore Small Pearl
(Thunderbolt - Ardmore Beauty)
Ardnasillagh Magnus (FR)
(MacDara - Blondie)
Ardnasillagh O'Flaherty (DEN/FIN)
(Bridge Boy - Ardnasillagh Treasure)
Ardravinna Andy
(Glyntown Paddy - Fenian Lass)
Ardravinna Finnain
(Glyntown Paddy - Fenian Lass)
Ashfield Alex
(Clonkeehan Auratum - Lambay Starry Eyes)
Ashfield Bobby Sparrow
(Carna Bobby - Wise Sparrow)
Ashfield Festy
(Granard Storm - Atlantic Surf)

Ashfield Gypsy (FR)
(Thunderbolt - Orphan Dolly)
Ashfield Judy
(Carna Bobby - Jump the Wall)
Ashfield Lucky Tim (FR)
(Killyreagh Kim - Lormount)
Ashfield Mactalla
(Ormond Oliver - Ashfield Lor Sparrow)
Ashfield Rolling Rock (FR)
(Thunderbolt - Jump the Wall)
Atalanta
(Atlantic Storm - Wild Rebel)
Atlantic Cliff (HOL)
(Dangan Dun - Hillswood Helena)
Atlantic Curragh (HOL)
(Dun Aengus - Atlantic Surf)
Atlantic Gael (HOL)
(Ben Lettery - Grey Dun)
Atlantic Mist
(Carna Bobby - Callowfeenish Dolly II)
Atlantic Noreen (SWE)
(Carna Dun - Careless Nora)
Atlantic Storm
(MacDara - Flash Girl)
Aughnish Dolly (GER)
(Ardilaun - Cashel Doll)
Aura Dun (USA)
(Clonkeehan Auratum - Early Grey)
Aylesland Silver Velvet
(Silver Snow - Wester Wind)

Ballinaboy Eamonn (USA)
(Loobeen Larry - Ocean Bambi)
Ballinaboy Jonquil
(Abbeyleix Owen - Ocean Bambi)
Ballinahinch Gold
(Abbeyleix Owen - Rose Garland II)
Ballintemple Dun Rambler
(Carna Dun - Moonlight Rambler)
Ballydonagh Bobby
(Carna Bobby - Glen Fanny)
Ballydonagh Deirdre (HOL)
(Carna Bobby - Cashel Kate)
Ballydonagh Kate
(MacDara - Cashel Kate)

Ballydonagh Rob
(Carna Bobby - Cashel Kate)
Ballydonnellan Grey
(Cilciarain - Rosaveel Grey)
Banba Chestnut (HOL)
(Island King - Easter Chestnut)
Bar Bar A's Kilkerrin Paddy (USA)
(Georgia's Mister Irish - Morning Mist)
Bardsey Flora
(Whitwell Cockade - Tubber Posy)
Bardsey Zodiac
(Whitwell Cockade - Holton Flash Favour)
Barna Honey
(Paddy's Heaven - Furbough Heather)
Barwidgee Errif (NZ)
(Connaught of Millfields - Barwidgee Frederika)
Barwidgee Fidler (AUS)
(Tulira Colman - September Song of Millfields)
Bay Lass
(Garryhinch Prince - Brown Lady)
Beal Tra
(Atlantic Storm - Truska Glory)
Bell Cazo (SWE)
(Brantshammer Bell Dara - Castanje)
Ben Lettery
(MacDara - Swanee River)
Bengt (GER)
(Big Ben - Lady Lifty)
Billy
(Catamaran - Valley Rose Marie)
Black Paddy
(CPS 8 - no breeding)
Bobby Brown (USA)
(Abbeyleix Ri - Brown Bird)
Bog Oak
(CPS 287 - no breeding)
Boogie Woogie (FR)
(Bridge Boy - Betsy Bay)
Breath of Spring
(MacDara - Frosty Winter)
Breidin (SWE)
(MacDara - Ballydonellan Grey)
Bridge Boy
(Tooreen Ross - Irene)
Bright Pearl (GER)
(Flagrant - Blossom Honey)
Brilliant Dawn
(Mervyn Kingsmill - Emlagh Dawn)
Brown Alice
(Dun Lorenzo - Lettershask)
Brown Lady
(Abbeyleix Owen - Trixie)
Brunskaer Marguerite (FIN)
(Oxenholm M.A.S.H. - Maysie)
Brunskaer Mars (FIN)
(Hazy Dawn - Maysie)
Bunowen Castle Lady
(Mervyn Kingsmill - Power Station)
Burrishoole Sandy
(Carna Bobby - Westport Sandy)

Cailin Ciuin
(Abbeyleix Owen - Wireless Wave)
Calla Boy
(Dun Aengus - Sliabh na mBan II)
Calla Lass (HOL)
(Paddy's Heaven - Grey Hop II)
Calla Rebel
(Rebel - Calla Roan)
Callowfeenish Beauty
(Clonjoy - Sliabh na mBan II)
Callowfeenish Colleen
(Ben Lettery - Callowfeenish Beauty)
Callowfeenish Dolly II
(Carna Dun - Callowfeenish Dolly)
Callowfeenish Mairtin
(Abbeyleix Owen - Wireless Wave)
Callowfeenish Pride
(Carna Dun - Rusheen Pride)
Callowfeenish Star
(Ashfield Festy - Homeward Bound II)
Callowfeenish Thunder (USA)
(Thunderbolt - Callowfeenish Colleen)
Camlin Capella
(Carna Dun - Camlin Cilla)
Camlin Cicada
(Carna Dun - Camlin Cilla)
Camlin Cilla
(Gil - Ciro)
Camlin Cirrus
(Gil - Ciro)
Camus Harp (FR)
(Clonjoy - Irish Harp)
Canal Cormac
(Thunderbolt - Rambling Home)
Canningvale Irish Dusk (NZ)
(Domo Cavallo Benjamin - Irish Mist)
Cannon Ball
(CPS 1 - no breeding)
Canrower Lass
(Carna Bobby - Speculation)
Careless Nora
(Clonkeehan Auratum - Brown Alice)
Carna Bobby
(Gil - Carna Dolly)
Carna Dolly
(CPS 442 - no breeding)
Carna Dun
(Little Heaven [T.B.] - Double Dun)
Carna Gold (HOL)
(Carna Dun - Carrabaun Colleen)
Carra Dinnish
(Sarsfield - Dimple)
Carrabaun Boy
(MacDuff - Anne Maree)
Carraway (GER)
(Carrabaun Boy - Canrower Jacqueline)
Cashel Kate
(Tully Lad or Carna Dun)
Cashel Lass
(CPS 11 - no breeding)
Castle Dame
(Mervyn Kingsmill - Village Star)
Castle Park
(Abbeyleix Owen - Trabane Castle)
Catamaran
(Murphy Rebel - Miss Charger)

Ceilog
(Clonkeehan Auratum - Pollnaclough Grey)
Celebration Cream
(Marble - Moycullen Mite)
Celtic Truska Prince
(Clonkeehan Auratum - Rebel Colleen)
Charlie
(CPS 2 - no breeding)
Charm Bracelet
(Clonkeehan Auratum - Charm)
Checkpoint Charlie
(Camlin Cicada - Castletown Lady)
Cheviot Bracken
(Smokey Shane - Celebration Cream)
Cheviot Bryony
(Windy Cove Ranger - Celebration Cream)
Cheviot Campion
(Smokey Shane - Celebration Cream)
Cheviot Clover
(Millfields Commodore - Celebration Cream)
Cheviot Crocus
(Island Duke - Celebration Cream)
Cheviot Midas (SWI)
(Smokey Shane - Celebration Cream)
Chiltern Arctic Venture (HOL)
(Scottish Venture [T.B.] - Arctic Moon)
Chiltern Cameo
(Truska Pimpernel - Chiltern Gazelle)
Chiltern Celtic Dancer
(Island Duke - Arctic Moon)
Chiltern Colm (USA)
(Corrib - Chiltern Gazelle)
Chiltern Copa (USA)
(Chiltern Curlew - Common Market)
Chiltern Curlew
(Island Duke - Arctic Moon)
Chiltern Dragonfly
(Silver Snow - Arctic Moon)
Chiltern Gazelle
(Island Duke - Arctic Moon)
Chiltern Humdinger
(Island Duke - Lune Star)
Chiltern Manikin
(Island Duke - Arctic Moon)
Chiltern Quota
(Chiltern Manikin - Common Market)
Chiltern Royalty
(Island Duke - Common Market)
Chiltern Saffron (AUS)
(Island Duke - Arctic Moon)
Chiltern Sunsilk (NZ)
(Island Duke - Arctic Moon)
Chiltons Way (NZ)
(Chiltern Sunsilk - Ridgeway Gem)
Cilciarain
(Gil - Irish Beauty)
Ciro
(Silver Pearl - Swallow II)
Clare Charger
(Mervyn Storm - Wild Judy)
Clifden Charm
(Doon Paddy - Clifden Molly)
Clifden Molly
(Cilciarain - Knock Molly)
Clifden Spirit
(Not known - Gurtroomagh Molly)
Clonjoy
(Clonkeehan Auratum - Joyce Grey)
Clonkeehan Archer (AUS)
(Strongbow - Camlin Cilla)
Clonkeehan Auratum
(Naseel [Arab] - Western Lily)
Clonkeehan Fuchsia (AUS)
(Camlin Cicada - Nipper of Calla)
Clonkeehan Kittiwake (HOL)
(Strongbow - Clonkeehan Fuchsia)
Clonkeehan Mayflower
(Naseel [Arab] - Rose of Barna)
Clonkeehan Nimbus
(Camlin Cicada - Dark Winter)
Clonkeehan Tiger Lily
(MacDara - Clonkeehan Easter Lily)
Clonkeehan Whitethorn
(Camlin Cicada - Clonkeehan Mayflower)
Clonkeehan Yeoman
(Clonkeehan Bowman - Camlin Cilla)
Clough Rebel
(Rebel - Derrada Fanny)
Cnocmordain Lass
(CPS 340 - no breeding)
Cocum Coppelia (USA)
(Atlantic Swirl - Grayswood Doon Bridge)
Cocum Jackpot (DEN)
(Cocum Hawkstone - Cocum Rebel Dolly)
Cocum Thunder Boy (FR)
(Thunderbolt - Cocum Bella Donna)
Common Market
(Rebel Wind - Free Trade)
Connemara Boy
(CPS 9 - no breeding)
Connemara Park Judy (AUS)
(Connemara Park Seamus - Ardan)
Connemara Park Michael (NZ)
(Island King - Ardan)
Connemara Park Richard (AUS)
(Connemara Park Seamus - Ardan)
Connemara Park Roy (NZ)
(Connemara Park Kenneth - Renvyle Rebel)
Connemara Park Wilfred (AUS)
(Clonkeehan Archer - Easter Peak)
Coolfin Beauty
(Carna Bobby - Grey Hop II)
Coomel Kestrel (AUS)
(Inis O'Hara - Little Lark of Abbeyleix)
Coosheen Alanna
(Tiger Gil - Tooreen Queen)
Coosheen Finn
(Carna Bobby - Finola of Leam)
Coosheen Nutmeg (GER)
(Coosheen Finn - Ganty Nutmeg)
Coosheen Petronella
(Carna Dun - Cregg Lassie)
Coral Misty
(Murphy Rebel - Gurteen Misty)
Coral Prince
(Murphy Rebel - Snowdrops Surprise)
Coral Star
(Cuchulainn - Dolan Lucky Star)

Coral Strand
(Doilin Star - Tishmane Orla)
Crackerjack (G)
(Thunderbolt - Woodlands Rover)
Creganna Winter
(Winter [T.B.] - Creganna Peggy)
Cregg Hero (SWE)
(Bridge Boy - Cregg Lassie)
Cregmore Colm (AUS)
(Tulira Rambler - Our Smokey)
Cregmore Dun (AUS)
(Tooreen Ross - Smokey)
Cregmore Kiaran
(Kilgreaney Lad - Lucky Smokey)
Crockaun Connemara (SAF)
(The Fugitive - Silver Lace)
Crusheen Connie
(Corbally Con - Nansin Bhan II)
Cuach na Coille
(CPS 236 - no breeding)
Cuach Og
(Gil - Cuach na Coille)
Cuaifeach
(Abbeyleix Owen - Breath of Wind)
Cuchulainn
(Abbeyleix Owen - Sparrow Hill)
Cushatrough Lass
(Farravane Boy or Carna Dun - Silver Rebel of Leam)

Dairina
(MacDara - Daut)
Dancing Spanner
(Little Heaven [T.B.] - no dam named)
Dangan Dun
(Rebel Wind - Breath of Spring)
Davy D.
(Killyreagh Kim - Diamonds Fancy)
Dawn Mist
(Ormond Oliver - Leam Little Lady)
Derry Boy
(Connemara Boy - Derry Girl)
Diana Van Graaf Janshof (HOL)
(Clonkeehan Yeoman - Ballydonagh Deirdre)
Dimple
(Clonjoy - Leam Darling)
Dimples
(Kilgreaney Lad - Kilcloghan Lady)
Domo Cavallo Praize (AUS)
(Connemara Park Peter - Kirtling Haze)
Doon Lily
(Noble Star - Molly Bawn II)
Doon Paddy
(Carna Dun - Doon Lily)
Dooneen Castle
(Mervyn Kingsmill - Dooneen Star)
Dooneen Grey
(Mervyn Storm - Rock of Cashel)
Downhill Dewdrop (FIN)
(Kasta Scaramouche - Aramara Audrey Grey)
Drimeen Dun
(Dun Lorenzo - Errislannan Blackie)
Dun Aengus
(Dun Lorenzo - Rose of Killola)
Dun Lorenzo
(MacDara - Draighneann Don)
Dun Orphan
(Dun Lorenzo - Orphan Maid)
Dun Sparrow
(Carna Dun - Wise Sparrow)
Dunlewey Sonny Jim
(Rebel Wind - Corn Flake)

Early Sunrise (FR)
(Dun Lorenzo - Bally Sunrise)
Easter Glory
(Carna Bobby - Leagaun Bay)
Easter Queen (FR)
(Island Duke - Heavenly Days)
Easter Song
(Minstrel Boy - Missel Thrush)
Easter Thorn (FR)
(Carna Dun - Blackthorn)
Errigal Prince
(Clonjoy - Errigal Princess)
Errigal Princess
(Tiger Gil - Furbough Lass)
Errisbeg Dolly
(Abbeyleix Owen - Errisbeg Rose)
Errisbeg Rose
(Carna Dun - Dolan Rose)
Errislannan Blackie
(Silver Pearl - no dam named)
Errislannan Coltsfoot
(MacDara - Drimeen Dun)
Errislannan Daisy
(Doon Paddy - Drimeen Dun)
Espe Etta (FIN)
(Cocum Tiger Moth - Espe Aletta)
Eudene Moonstone
(Ardravinna Andy - Pillagh Grey Pearl)
Eva de Ravary (FR)
(Island Earl - Screen Lady)
Evening Pearl
(Silver Pearl - Lettershask)

Fairy Hill Lass
(Ashfield Festy - Harvest Home)
Fairy Legend
(Carna Dun - Little Nell)
Far Above's Katherine (U.S.A.)
(Far Above's Shannon - Lambay Maeve)
Ferdia
(Murphy Rebel - Wireless Wave)
Finney Master (FR/SWE)
(Carna Dun - Winter Roche)
Finola of Leam
(Lavalley Rebel - Teresa of Leam)
Fionnuala
(Atlantic Storm - Finola of Leam)
Flagrant (GER)
(Flashy Lad - Una's Friend)
Floor Colleen (HOL)
(Atlantic Storm - Dillons Bay)
Florence Nightingale
(Cilciarain - Florence)
Fort Hazel
(Marble - Fort Sapphire)

Fort Holly
(Clonkeehan Auratum - Irene Grey)
Fort Lambay
(Marble - Irene Grey)
Fort Sapphire (FR)
(Glenarde - Moycullen Mite)
Fort Viola
(Carna Bobby - Fort Pansy)
Fortuna (GER)
(Little Lord - Flash Princess)
Free Trade
(Gleann Aoibhinn - Faulkera Star)
Fudge IV
(Clonkeehan Auratum - Derryeighter Bay)
Funny Girl
(Cream Cracker - Lettery Tempest)
Furbough Heather
(Cilciarain - Glann Chestnut)
Furbough Lass
(Skibbereen [I.D.] - Furbough Peggy)

Gael Linn (USA)
(MacDara - Cloon Bay)
Ganty Gina
(Doon Paddy - Ganty Jane II)
Garrafin
(Rebel Wind - Bobby's Pride)
Garryhinch Milrace (FR)
(Abbeyleix Fionn - Kilbracken Queen)
Gil
(Inchagoill Laddie - Golden Gleam)
Gillian
(Killyreagh Kim - Callowfeenish Beauty)
Gina of Ganty (SWE)
(Paddy's Heaven - Ganty Jane)
Gina (GER)
(Golden Rock - Aughnish Dolly)
Glann Fireball
(Jimmy's Fireball - Charm Bracelet)
Glaskopf Golden Merlin (GER)
(Golden Dan - Maisie)
Glen Nelly
(Gil - Nelly Carroll)
Glenarde (HOL)
(MacDara - Ballinahown Grey)
Glencara Pet
(Dun Lorenzo - Kilkerrin Lass)
Glencaugh Star
(Inver Rebel - Lor Lark)
Glendunn (FR)
(Lambay Fireball - Caroline)
Gloves Misty
(Carna Bobby - Grey Hop II)
Glyntown Paddy
(Doon Paddy - Truska Colleen)
Gold Fort (FR)
(Marble - Fort Silver)
Gold Ring
(Inver Rebel - Gold Flake)
Golden Dan (GER)
(Ben Lettery - Ballydonnellan Grey)
Golden Discovery
(Mervyn Kingsmill - Power Station)

Gonzales (GER)
(Granger - Sky Girl)
Good Heaven (G 25)
(Carna Dun - Heather Grey)
Gorteen Boy
(Silver Rocket - Gloves Misty)
Granard Storm
(Atlantic Storm - Martins Pride)
Greaney Rebel
(Inver Rebel - Nansin Ban)
Gregory (SWE)
(Cregg Hero - no dam named)
Grey Aoife
(Fermoyle Boy - Fair Aoife)
Grey Crayon
(Ardilaun - Rinso)
Grey Dawn Lassie
(Inchagoill Laddie - Water's Grey)
Grey Hop
(CPS 1674 - no breeding)
Grey Hop II
(Mervyn Storm - Grey Hop)
Grey Rock Star
(Abbeyleix Owen - Breath of Wind)
Gurtroomagh Molly
(Murphy Rebel - Clifden Molly)

Happy Rebel
(Inver Rebel - Happy Patricia)
Haughmond Danielle
(Chiltern Manikin - Inver Heather)
Hazy Dawn
(Dale Haze - Castle Park)
Heather
(CPS 158 - no breeding)
Heather Bell
(CPS 15 - no breeding)
Heather Cuckoo
(Mervyn Kingsmill - Wise Cuckoo)
Heather Mixture of Abbeyleix
(Clonkeehan Auratum - Calla Brown)
Hideaways Greystone Alex (USA)
(Hideaways Erin Smithereen - Lynfields Proud Hope)
Hill Country
(Inver Rebel - Callowfeenish Bay)
Home Lover
(Clonjoy - Homeward Bound)
Homeward Bound
(Rebel Wind - Jump the Wall)
Homeward Bound II
(Thunderbolt - Homeward Bound)
Honey Belle
(Creganna Winter - Dun Belle)
Hummelbro Ballerina (SWE/FIN)
(Hummelbro Daf - Errigal Beauty)

Ice Cream
(Toombeola - Village Belle)
Idenoir (FR)
(Windy Cove Ranger - Little Cashel Hill)
Inchagoill Hero
(Rebel - Mynish Lass)

Inchagoill Laddie
(Rebel - Dooyher Lass)
Inchagoill Star
(Noble Star - Athry Maid)
Indra Rebel (FRA)
(Rebel Wind - Ashfield Surf)
Inis O'Hara (AUS)
(Thunderbolt - Forest Flower)
Inver Rebel
(Lavalley Rebel - Inver Bridge)
Inverin Rosebud
(Ben Lettery - Primrose Path)
Irene
(Inchagoill Laddie - Silver Grey)
Irene II
(Farravane Boy - Irene)
Irish Darling
(Dun Lorenzo - Irish Lass)
Irish Harp
(Inver Rebel - Grey Hop)
Irish Lass
(Noble Star - Bog Oak)
Irish Town Beauty
(Abbeyleix Fionn - Irishtown Diana)
Island Duke
(Clonjoy - Grey Dawn Lassie)
Island Earl (FR)
(Carna Dun - Helpmate)
Island King (AUS)
(Carna Bobby - Doon Lass)
Island Lad
(Kilgreaney Lad - Broadstone Lass)
Island Storm
(Milford Hurricane - Lovely Deirdre)

Jennifer Rose
(Mervyn Kingsmill - Sukie)
Jessica (GER)
(Carrabaun Boy - Canrower Jacqueline)
Jimmy's Fireball
(Clonkeehan Nimbus - Bobby's Girl)
Jolly Roger
(MacDara - Callowfeenish Dolly II)
Julie
(Carna Bobby - Windy Dun)
Jump the Wall
(Calla Rebel - Cuach Og)

Kasta Scaramouche (SWE/FIN)
(Blue Chief - Erins Own)
Katie (AUS)
(Dun Aengus - Slieve Winter)
Kilduffahoo
(Sarsfield - Tir Na n-Og)
Kildunn Mayflower
(Carrabaun Boy - Cashel Guide)
Kilgreaney Lad
(Carna Dun - Lissoughter Dolly)
Kilkerrin Cream
(Connemara Boy - no dam named)
Kilkerrin Surf
(Thunderbolt - Ashfield Surf)
Killyreagh Jane
(The Fugitive - Ballydonagh Deirdre)
Killyreagh Kim
(Carna Bobby - Ballydonagh Kate)
Kimble (SWE)
(Dun Aengus - Inver Colleen)
Kingstown
(Clonkeehan Auratum - Kingstown Grey)
Kingstown Molly
(Killyreagh Kim - Park Princess)
Kingstown Swallow (HOL)
(MacDara - Kingstown Grey)
Kinvara Hop
(Paddy's Heaven - Grey Hop)
Kits Bay
(Jimmy's Fireball - Charm Bracelet)
Knock Molly
(CPS 1289 - no breeding)
Knockabout Lady
(Dun Aengus - Home Lover)
Knockadreen Holiday Cliff (FR)
(Ballintemple Dun Rambler - Bobby's Girl)
Knockranny Ruby
(Clonkeehan Auratum - Star of Fahy)

Lake View Rambler
(Tulira Rambler - Our Smokey)
Lambay Bobby
(Carna Bobby - Ganty Jane)
Lambay Fireball (FR)
(Camlin Cicada - Clonkeehan Easter Lily)
Lambay Inver (SWE)
(Camlin Cicada - Clonkeehan Inverbeg)
Lambay Kate
(Knockranny Ruby - Oldtown Waterfall)
Lambay Rebel (DEN)
(MacDara - Moonshine)
Lambay Star
(Carna Dun - Gay Lass)
Lavalley Rebel
(Rebel - Derrada Fanny)
Leam Darling
(Bridge Boy - Irish Darling)
Leam Lassie
(Tooreen Laddie - Irish Lass)
Leam Little Lady
(Rebel Wind - Leam Lassie)
Lehid Colleen
(Mervyn Kingsmill - Lehid Lovely Princess)
Lehid Ladybelle
(Abbeyleix Owen - Village Belle)
Lehid Linda
(Killyreagh Kim - Lehid Wild Wind)
Lehid Lovely Princess
(Killyreagh Kim - Lehid Star)
Lehid Rose
(Clough Droighnean - Little Nell)
Leppa Joy (FIN)
(Clonjoy - Leppa Grey)
Lettershask
(Noble Star - no dam named)
Lily of the Valley
(Skryne Bright Cloud - Ardmore Small Pearl)
Lissoughter Cora
(Ormond Oliver - Lissoughter Laura)

Lissoughter Dolly
(Carna Bobby - Ruby)
Lissoughter Laura
(Clonjoy - Lissoughter Dolly)
Little Joe
(Carna Dun - Cregg Lassie)
Little Nell
(Silver Pearl - no dam named)
Little Rascal (BEL)
(Seafield Silver Rogue - Trabane Castle)
Lonsdale Windfall (AUS)
(Connemara Park Michael - Windfall)
Loobeen Larry
(Killyreagh Kim - Loobeen Meg)
Loobeen Meg
(Thunderbolt - Loobeen Lily)
Lough Easkey
(Doon Paddy - Rambling Colleen)
Lougheera
(CPS 228 - no breeding)
Lucky Smokey
(Bridge Boy - Smokey)
Lune Star
(Bridge Boy - Arctic Moon)
Lynch Law
(Marconi - Dublin Grey)

Maam Hill
(Dun Aengus - Bay Lady)
MacDara
(Dun Lorenzo - Dolan Rose)
Macduff (FR)
(Dun Aengus - Callowfeenish Pride)
Macnamara
(MacDara - Nora)
Majestic Lass
(Tooreen Ross - Molly Ban)
Mancho Van Graaf Janshof (HOL)
(Sticky - Clonkeehan Kittiwake)
Marble (DEN)
(Rebel Wind - Callowfeenish Dolly II)
Marco Polo of Clonkeehan
(Bridge Boy - Canrower Lass)
Marconi (USA)
(MacDara - Wireless Wave)
Marena (GER)
(May Prince - Mary)
Marvel Van Graaf Janshof (HOL)
(Carna Gold - Ballydonagh Deirdre)
Matchmakers Lad
(Mervyn Kingsmill - Foreglass Matchmaker)
Merry Dolly
(Doon Paddy - Orphan Dolly)
Mervyn Blue Charm
(Abbeyleix Cypress - Windy Rebel)
Mervyn Invicta
(Camlin Cicada - Windy Rebel)
Mervyn Kingsmill
(Atlantic Cliff - Mervyn Blue Charm)
Mervyn Mayfly
(Camlin Cicada - Windy Rebel)
Mervyn Mistral
(Naseel [Arab] - Windy Rebel)
Mervyn Pookhaun
(The Admiral - Mervyn Nasim)
Mervyn Storm
(Camlin Cicada - Windy Rebel)
Midnite Marcus
(Rustle of Spring - Shipton Credhe)
Milady de Ravary (FR)
(Robber Boy - Screen Lady)
Milford Hurricane (SAF)
(Mervyn Bowsprit - Mervyn Silver)
Milford Wren
(Sarsfield - Mervyn Wren)
Millfields Commodore (AUS)
(Wisbridge Erinmore - Cornamona of Millfields)
Millfields Ensign (FIN)
(Wisbridge Erinmore - Gentle Breeze of Millfields)
Millie of Abbeyleix
(Carna Dun - Abbeyleix Molly)
Miss Charger
(Clare Charger - Story of Slievenamon)
Molly Ban
(Inver Rebel - no dam named)
Moonlight Rambler
(Dun Orphan - Cuach Og)
Moorland Juliet
(Murphy Rebel - Lucky Juliet)
Mountain Lad
(CPS 32 - no breeding)
Mountain Silverwing
(Rebel Wind - Slieve na mBan II)
Moy Hazy Cove
(Hazy Dawn - Windy Cove)
Moy Johnny's Pride
(The Kid - Castle Park)
Moy Village Master
(Finney Master - Village Dun)
Moycullen Mite
(Cilciarain - Grainne Mite)
Moyglare Samson
(Strongbow - Clonkeehan Winter)
Muffy
(Clonkeehan Auratum - Molly Conneely)
Mullaghmore Siskin
(Mervyn Invicta - Ballydonagh Deirdre)
Murphy Rebel
(Rebel Wind - Muffy)
Murrisk
(MacDara - Grey Girl)

Nansin Ban
(Camlin Cirrus - Clynagh Rose)
Nansin Bhan II
(Rebel Wind - Majestic Lass)
Napoleon
(Finney Master - Celebration Cream)
New Beginnings
(Quiz Master - Tulira Colleen)
Newbridge Oliver
(Ormond Oliver - Callowfeenish Colleen)
Nimble Fingers
(Clonkeehan Auratum - Lorenzo Lass)
Nipper of Calla
(Tiger Gill - Calla Lass)

Noble Star
(CPS 17 - Black Paddy - no dam named)
Noreen Grey
(Gael Linn - Rebel Noreen)
Nougatine Melody (FR)
(Grange Finn Sparrow - Carrabaun Cracker)
Nyton Village Master
(Rosenaharley Macmaster - Grayswood Village Cuckoo)
Ocean Bambi
(Carna Dun - Banba Chestnut)
Ocean Seames (HOL/GER)
(Doon Paddy - Banba Chestnut)
Ocean Sky
(Errigal Prince - Muffy)
Ormond Oliver
(MacDara - Ashfield Judy)
Orphan Dolly
(Dun Orphan - Carna Dolly)
Our Smokey
(Aura Dun - Smokey)
Owens Lady
(Abbeyleix Owen - Tullabrick Jenny)
Oxenholm Bardot (DEN)
(Rosenaharley Mac Milo - Brigitte)
Oxenholm Mandy (DEN)
(Marble - Sally Maree)
Oxenholm O'Connell (DEN)
(Ardnasillagh O'Flaherty - Holens Smut)

Paddy's Heaven
(Carna Bobby - Irene II)
Party Gay Girl
(Ben Lettery - Homeward Bound)
Patsy Fagan (G)
(*Tooreen Ross - Smokey*)
Pigeon
(Ashfield Alex - Pigeon on the Strand)
Pigeon on the Strand
(Abbeyleix Owen - Fairy Legend)
Pilgrim
(Abbeyleix Owen - Lehid Wild Wind)
Pillagh Beauty
(Cilciarain - no dam named)
Pillagh Grey Pearl
(Marble - Pillagh Beauty)
Pillagh Star
(Carna Bobby - Ever Ready)
Power Station
(Carna Dun - Truska Brownie)

Quiz Master
(Atlantic Storm - Evening Stroller)

Rambling Colleen
(Gil - Rockfield Girl)
Rambling Home
(Carna Bobby - Ganty's Pride)
Randam du Bailly (FR)
(Harris de la Dive - Fancy Friend)
Rebel Lass
(Dun Orphan - Doirin Lass)
Rebel Wind
(Inver Rebel - Windy)
Red Baroness (DEN)
(Island Duke - Aylesland Silver Treasure)
Retreat
(CPS 320 - no breeding)
Robber Boy (FR)
(Clonjoy - Rinso)
Robe Mist
(Milford Hard Times - Kilbride Star)
Robin Hood
(Skyrne Bright Cloud - Village Grey)
Rocky (USA)
(Cuchulainn - Silver Fort)
Roobeen Boy
(Clonjoy - Grey Steel)
Rosaveel Storm
(Mervyn Storm - Queen of the Hills)
Rose of Barna
(Man of Barna - White Nell)
Rosenaharley Laurin
(Rosenaharley Cormac - Leam Laurella)
Rosenaharley Rossleague
(Atlantic Swirl - Rosenaharley Rosmara)
Rosmuck Master
(Clonjoy - White Lor)
Roundmount
(Carna Dun - Lormount)
Roundstone Oscar (CAN)
(Rebel Wind - Dancing Spanner)
Roundstone River (FR)
(Rebel Wind - Dancing Spanner)
Ruby
(Mayboy [I.D.] - Retreat)
Rustle of Spring
(Celtic Truska Prince - Llandegvedd Bean Si)
Rynn Richard
(Ardravinna Andy - Fuchsia)

Safir de Beauchamp (FR)
(Abbeyleix Apollo - Camus Harp)
Sarsfield
(Killyreagh Kim - Wise Cuckoo)
Scarlett (GER)
(Golden Dan - Scraugh Colleen)
Screen Lady (FR)
(Rebel Wind - Miss Dunheath)
Seafield Silver Rogue
(Silver Rocket - April Anita)
Silva
(CPS 1102 - no breeding)
Silver Duke
(Island Duke - Silver Lass)
Silver Finn
(Ardilaun - Finola of Leam)
Silver Fort
(Rory Ruadh - Fort Hazel)
Silver Lass
(Errigal Prince - Florence Nightingale)
Silver Pearl
(CPS 18 - believed by John Quirke - Bess)
Silver Rocket
(Doon Paddy - Slieve Winter)

Silver Sparrow (FR)
(Abbeyleix Owen - Sparrow Hill)
Skryne Bright Cloud
(Clonkeehan Nimbus - Aughris Bright)
Sky Line (CAN)
(Pegasus - dam unknown)
Sliabh na mBan
(CPS 227 - no breeding)
Sliabh na mBan II
(Calla Rebel - Sliabh na mBan)
Sliabh na mBan Cailin
(Abbeyleix Owen - Sliabh na mBan Peggy)
Sliabh na mBan Peggy
(Killyreagh Kim - Mountain Silverwing)
Slieve Winter
(Creganna Winter - Slieve na mBan)
Slisenoir
(*Cuchulainn - Milford Wren*)
Smokey
(Connemara Boy - Kilkerrin Cream)
Smokey Duncan
(Westside Frank - Smokey Jane Grey)
Smokey Fireman
(Westside Frank - Smokey Latecomer)
Smokey Jane Grey
(Kilgreaney Lad - Our Smokey)
Smokey Latecomer
(Kilgreaney Lad - Our Smokey)
Smokey Playboy
(Tulira Mairtin - Our Smokey)
Smokey Shane
(Kilgreaney Lad - Lucky Smokey)
Snoopy V.D. Leliard (HOL)
(Calmore Dew - Sunny)
Snowdrops Surprise
(Ben Lettery - Snowdrop)
Snowflake
(Creg Coneera - Heath Winter)
Solitude
(Clonkeehan Auratum - Loobeen Queen)
Song of Regine
(The Fugitive - Easter Song)
Spanish Arch
(Marconi - Kilkerrin Grey)
Spratt Lady
(Kilgreaney Lad - Kilcloghan Lady)
Star of Fahy
(Dun Lorenzo - Molly Conneely)
Sticky (HOL)
(Carna Bobby - Cashel Kate)
Streamstown Larry
(Mervyn Kingsmill - Muffeen)
Strike Action (HOL)
(Killyreagh Kim - Power Station)
Strongbow
(Calla Rebel - An tSailchuach)
Sukie
(Carna Bobby - Jennifers Girl)
Suzan Van Graaf Janshof
(Clonkeehan Yeoman - Ballydonagh Deirdre)
Swallow II
(CPS 473 - no breeding)
Sweet Lass
(Abbeyleix Owen - Woodlands Rover)
Swift Sally
(Killyreagh Kim - Careless Nora)
Sylvan Vale
(Tully Lad - Silva)

Tantallon Bobby
(Carna Bobby - Breath of Spring)
Tegelstrups Baroness (DEN)
(Oxenholm Hawkeye - Red Baroness)
The Admiral
(Carna Bobby - Cashel Kate)
The Fugitive
(Kimble - Loobeen Lily)
The Kid (ITY)
(Abbeyleix Fionn - Carrabeg Bonnet)
Thunder
(Tully Lad - Saint Kathleen)
Thunderball
(Canal Cormac (Dawn Mist)
Thunder Cloud
(Thunderbolt - Claddagh Queen)
Thunder Cloud (G.57)
(Doon Paddy - Smokey)
Thunder Storm
(Thunderbolt - Dooneen Grey)
Thunderbolt
(Thunder - Irene Grey)
Tiercel Witch Hazel
(Leam Bobby Finn - Wisbridge Silver Birch)
Tiger Gill
(Gil - Tiger Tank)
Toombeola
(Carna Dun - Silver Belle)
Tooreen Ross
(Tooreen Laddie - Wayfarer)
Tower Hill's Dareena (USA)
(Chiltern Copa - Cocum Coppelia)
Tower Hill's Wizard of Oz (USA)
(Chiltern Copa - Springledge Dublin Lass)
Trixie
(Rebel Wind - Keeraun Lassie)
Truska Colleen
(Inchagoill Laddie - Little Nell)
Truska Pimpernel
(Killyreagh Kim - Wise Colleen)
Tubber Posy
(Kimble - Happy Rebel)
Tulira Colman
(Bridge Boy - Julie)
Tulira Crackerjack (FR)
(Cregmore Colm - Tulira Charm)
Tulira Finn McCool
(Tulira Mairtin - Tulira Heather)
Tulira (G 115)
(Clonjoy - Smokey)
Tulira Guinness
(Tulira Mairtin - Mac's Bar)
Tulira Linnet
(Carna Bobby - Sandpiper)
Tulira Mairtin
(Tooreen Ross - Glen Nelly)
Tulira Mary Claire
(Tulira Mairtin - Noreen Grey)

Tulira Nimble Dick
(Tulira Mairtin - Noreen Grey)
Tulira Rambler
(Clonjoy - Truskey Grey)
Tulira Smasher (BEL)
(Tulira Mairtin - Noreen Grey)
Tully Blue
(Tully Grey - Blue Duster)
Tully Grey
(Tully Lad - Cait ni Dhuibhir)
Tully Lad
(Mountain Lad - Tully Beauty)
Turlough Beauty
(Ganty Champ - Shannagh Beauty)
Turlough Harp
(Abbeyleix Owen - May Harp)
Tynagh de Briace (FR)
(Nippon II - Nancy Kate)

Van De Arenbosch Frederique (HOL)
(Kimble - Banba Chestnut)
Vanesa (HOL)
(Paddy of Caherlistrane - April)
Venise Melody (FR)
(Idenoir - Nougatine Melody)
Vera
(Carna Dun - Caitin Aluinn)
Village Belle
(Calla Rebel - Village Swallow)
Village Boy
(Mervyn Kingsmill - Village Grey)
Village Colleen
(Mervyn Kingsmill - Village Star)
Village Grey
(Abbeyleix Owen - Village Belle)
Village Laura
(Thunderbolt - Village Star)
Village Star
(Killyreagh Kim - Village Belle)

Washlands Boann
(Silver Snow - Easter Island of Checkendon)
Waterfall Enniskillen (NZ)
(Connemara Park Roy - Blandings Kimberley)
Waterfall Moycullen (NZ)
(Connemara Park Roy - Blandings Kimberley)
Westport Sandy
(Paddy's Heaven - Woodlands Rover)
Westside Frank (GER)
(Bobby Brown - Ballydonagh Suzy)
Westside Fred
(Checkpoint Charlie - Ballydonagh Mary Anne)
White Granite (FR)
(Marble - Ganty Jane)
Whitewood Galway Bay (USA)
(Clonkeehan Auratum - Keanes Selection)
Wild West Rain
(Fair Justice - Gentle Dolly)
Willow
(Atlantic Storm - Callowfeenish Dolly II)
Windy
(Heather Bell - Furnace Lass)
Windy Day
(Killyreagh Kim - Windy Joy)
Windy Joy
(Clonjoy - Windy June)
Windy Rebel
(Lavalley Rebel - Windy)
Windycove
(Tooreen Ross - Windy Dun)
Windycove Ranger (FR)
(Carna Bobby - Windy Cove)
Winnie
(CPS 141 - no breeding)
Winnie Love
(Ashfield Festy - Winnie Dear)
Winnie the Pooh
(Doon Paddy - Winifred)
Winning Ways
(CPS 1172 - no breeding)
Winter Heather
(Creganna Winter - Heather Black)
Winter Mollie (USA)
(Creganna Winter - Cnocmordain Lass)
Winter Roche
(Creganna Winter - Dun Roche)
Wireless Wave
(Clonjoy - Callowfeenish Pride)
Wisbridge Erinmore
(Wisbridge Golden Rebel - Ayle Glory)
Wisbridge Golden Virginia
(Wisbridge Erinmore - Glencara Silver Star)
Wise Cuckoo
(Clonkeehan Auratum - Wise Sparrow)
Wise Sparrow
(Dun Lorenzo - Lor Sparrow)
Wistful Alice
(Clonkeehan Auratum - Brown Alice)
Woodham Silver Pearl
(Little Joe - Silver Lass)
Woodlands Rover
(Doon Paddy - no dam named)
Woodquay Fireman (FR)
(Jimmy's Fireball - Charm Bracelet)
Wychwood Grey Gull (AUS)
(Coomel Kestrel - Ardan)

Yarraman Park Henry (AUS)
(Connemara Park Billy - Katie)

Zet Dixie
(Killyreagh Kim - Zet Rebel)